HOW TO MELLIFY A CORPSE

And Other Human Stories of Ancient Science & Superstition

VICKI LEÓN

Walker and Company
New York

Published by Walker Publishing Company, Inc., New York

All papers used by Walker & Company are natural, recyclable products made from wood grown in well-managed forests. The manufacturing processes conform to the environmental regulations of the country of origin.

LIBRARY OF CONGRESS CATALOGING-IN-PUBLICATION DATA

León, Vicki.
How to mellify a corpse : and other human stories of ancient science & superstition / by Vicki León.—1st U.S. ed.
p. cm.
Includes bibliographical references and index.
ISBN 978-0-8027-1702-3 (alk. paper)
1. Science, Ancient—Miscellanea. I. Title.
Q124.95.L46 2010
509.3—dc22
2009044773

Visit Walker & Company's Web site at www.walkerbooks.com

First U.S. edition 2010

1 3 5 7 9 10 8 6 4 2

Designed and typeset by Suzanne Albertson
Printed in the United States of America by Worldcolor Fairfield

Praise for *How to Mellify a Corpse*

"...tly combines modern research with ancient lore to lift the lid on the classical world's weird and wonderful, ranging from solar fountains and surround-sound to lethal lipstick. Now I know how to predict the future while feeding chickens. Marvelous stuff!"

—RUTH DOWNIE, author of *Persona Non Grata: A Novel of the Roman Empire*

"Like a literary time machine, Vicki León's book plunks readers into the sandals of antiquity's greatest sages, soldiers, and kings. A captivating read from start to finish."

—ALAN HIRSHFELD, astrophysicist and author of *Eureka Man: The Life and Legacy of Archimedes*

"Vicki León's *How to Mellify a Corpse* is a scintillating compendium of ancient beliefs and practices, from magical thinking to proto-scientific inklings."

—ADRIENNE MAYOR, research scholar of classics and history of science at Stanford University and author of *The First Fossil Hunters* and *Greek Fire, Poison Arrows, and Scorpion Bombs*

"Suffering from the evil eye? Thinking of raising a pet eel? Look no further. With wit and insight, author Vicki León brings the wonders of the ancient world to light. From stench wars to funeral rites, León takes us on a fantastic journey to the origins of ideas and ideologies. A rollicking reference."

—STEPHANIE LILE, museum educator and author of *History Lab to Go!*

"León returns with another volume of fascinating miscellany, this time about the thought-world of the Greeks and Romans. Her lively anecdotes range from Stoics to stargazers, from ceremonial scapegoats to (mostly pseudo)scientists. In a world still poised between superstition and reason, León shows us that the examples of the ancients are more pertinent than you might think."

—STEVEN SAYLOR, author of *Empire: The Novel of Imperial Rome*

"Vicki León has done it again, turning a litany of ancient names into warts-and-all portraits of real philosophers, early scientists, and architects. With acerbic wit, she makes accessible the complex teachings of our icons from the deep past."

—DR. JACQUELINE WALDREN, Institute of Social and Cultural Anthropology and International Gender Studies at University of Oxford

"The entries' titles and subtitles of the book's entries are factual and often humorous. León is good at engineering hilarious incongruity with fractured idioms—"Measurements and money: The whole nine cubits," for instance. My favorite entries are "Acoustics: The first surround sound" and "Timekeeping: Calendar wars" because they made me curious about subjects new to me. Highly entertaining and intellectually stimulating."

—Caroline Hatton, Ph.D., analytical chemist and author of *The Night Olympic Team*

"Vicki León's penetrating account of the Greco-Roman world ranges from forward-looking science, technology, and philosophy to mind-boggling superstition. All of it brought to life through a huge cast of real people, from the renowned to the undeservedly obscure, and told with style, humor, and energy. It's a perfect combination: the ancient world seen through perceptive modern eyes."

—Stephen Moorbath, professor emeritus at University of Oxford

"This conversational book about long-ago people, places, and processes is organized geographically rather than thematically, with alternating emphasis to keep us engaged. Given my background, I was especially interested in the entries on geography and geology. A picture may be worth a thousand words, but Vicki's book delivers a thousand years of fascinating anecdotes."

—Jerry Clark, college geography teacher and NASA space researcher, retired

"A thought-provoking read. The facts are fun and informative; exactly what Latin teachers (and students) love to share. Furthermore, Vicki plays with words so eloquently and lightly that she draws in readers. The connections she makes to the ancient world through science will keep students delving deeper. I will be adding elements of her book to my class as a spice to keep them savoring the classics."

—Brian Geffre, high school Latin and English teacher and sponsor of Shanley Junior Classical League

This book is dedicated to

Kieran Conroy, whose eager young curiosity thrills me,

Sandee Ogren, whose wisdom, wit, and grit inspires me,

and Stephen Moorbath, whose talent for science

and tomfoolery slays me

CONTENTS

SECTION II

Greece & the Greek Islands 45

SECTION III

Asia Minor & the Middle East 91

SECTION V

Italy & Sicily 197

SECTION VI

North Africa & Mesopotamia 249

Dear reader,

You're beginning a field trip to a long-ago world filled with science, superstition, and the folks who believed in both. Because its ground rules were quite different from our own, here's a quick tutorial.

Names. Ancient Greeks went by one name plus birthplace, as in Diogenes of Sinope. In this book, individuals are identified by their best known names (and spellings): Cicero, Marc Antony, Cleopatra, Archimedes, Herodotus. Elite Romans (and even some freedmen) had three-part names but I've spared you most of them. The glaring exception? Rome's first imperial ruler. Born Gaius Octavius, the grand-nephew of Julius Caesar, he later became Caesar's adoptive son. During his forty-one-year run, he accumulated a lengthy string of names and titles. In the interests of author sanity, I refer to him as Octavian before he became top dog, and (Emperor) Octavian Augustus after 27 B.C.

My username—which one? I have twenty-three.

Places. Although travel was difficult, Greeks and Romans logged an astonishing number of miles: job transfers, education abroad, marriage migrations. At times their zigs and zags meant rotten luck: captured by pirates, exiled for political reasons, or fleeing war. Since this book goes back as far as 700 B.C. and includes events as late as A.D. 300, the maps with each section will help you keep track of these Greco-Roman cultures.

The term "Greco-Roman" is a misleading convenience. Early on, the Greeks colonized, planting their city-states in Italy, North Africa, Spain, France, and around the Black Sea. Greek-speaking city-states also flourished at the eastern end of the Mediterannean, a region everyone called "Asia Minor" back then. Despite the name, each *polis* or city-state was more than an urban center.

Some city-states, Sparta for one, had huge amounts of land; others had only a crumb on which to raise crops. The most populous polis was Athens, which included the countryside of Attica around it.

Rome arose as a small city. After it devoured the local Etruscans, Rome got a taste for expansion. By 146 B.C., it had "freed" all the Greeks by swallowing them. Two decades after Julius Caesar was assassinated and his heir Octavian took the reins of power, the imperial centuries began, with the city of Rome as its capital. Lands conquered by the Romans paid tribute and became known as provinces.

Whether founded by Greeks or established by Romans, the fortunes of individual cities rose and fell. Some had real staying power, especially Athens and its cultural rival, Syracuse of Sicily; and Rome and its economic rival, Antioch of Syria. Pergamum, Ephesus, and Alexandria the inventive also enjoyed many centuries of influence. Others from Corinth to Carthage to Tyre had briefer yet substantial lifespans.

Look at it philosophically—at least we live near the Olympic Games. We could be stuck in Greek North Africa.

Buying power. The translation of ancient currency to modern money is a tricky goal, since this book covers a millennium and tracks hundreds of locales, many of which produced their own coinage. As a rule of thumb, the Greek drachma and Roman denarius were roughly equivalent. For centuries, daily wages for many workers ran between one and three drachmas or denarii—enough to cover food, shelter, and the basics.

Time periods. Timekeeping, reconciling lunar and solar calendars, was a nightmarish problem for ancient societies. (Skeptical? See the entry called Calendar Wars). In this book, B.C. and A.D. dates are there largely to give you points of reference and comparison.

To learn more. Countless people, places, events, and topics in this book intertwine in interesting ways, so you'll find them cross-referenced in the index; look for the "See also" cues.

INTRODUCTION

During a meditative stroll around his home turf, the region of Turkey that the Greeks of old called Asia Minor, a keen-eyed thinker named Thales stumbled across naturally occurring magnets called lodestones. Experimenting, he discovered their ability to attract iron; back in 600 B.C., this amounted to headline news. Giving the world its earliest sound bite, he exclaimed, "Lodestone made the iron move—it has a soul!" With that statement, he rejected the prevailing belief about inexplicable events: that the gods must have done it. That took courage.

Yikes. Which way is Egypt? I'm late for my sage internship.

Thales spent his life inquiring into the animating principles of the universe, the deeper nature of matter. Like other Greek seekers, he embraced learning from more ancient cultures, studying geometry and astronomy with the Egyptian sages. With his newly won knowledge he was able to accurately predict a solar eclipse, forcing armies to cancel a perfectly good battle slated for that day. This insightful eccentric has been called the first Greek scientist.

In the same era, his class act was echoed by Pythagoras, who sought answers to the universe in numbers and in music. A Greek born on the island of Samos, Pythagoras chose to establish his community of three hundred like-minded geeks, male and female, in southern Italy. It would grow to include thousands of adherents, including his wife and daughters, and thrive for centuries.

These beginnings of real science coincided with the birth of Greek philosophy, literally the love of wisdom, a framework for contemplating the "what's it all about?" questions. No one used the Latin-derived word *scientist* yet. Instead, Thales and Pythagoras called themselves natural philosophers or *physicists,* from the Greek *physika,* meaning "to come into being."

Whatever moniker they chose, these lovers of wisdom had a feverish curiosity about the world, seen and unseen. Philosophers from Aristotle to Zeno developed strong opinions on natural phenomena, on morality, on what constituted the good life. Some, such as Democritus, Heraclitus, and Lucretius, explored the unseen, from germs to atoms. Others, including Empedocles and Theophrastus, did pioneering work in ecology, botany, climate, and evolution. Still others, such as Hipparchus and Anaxagoras, studied the heavens, predicting eclipses and meteor crashes, all while trying to square the circle for the first time.

These early inquirers weren't confined to what we think of as Greece, either. Their hometowns ranged from Pellas in Macedon to Lampsacus in Asia Minor, from Sinope around the Black Sea to communities near the Dead Sea. Syracuse on Sicily, which for five centuries rivaled Athens in size, wealth, and scientific brilliance, boasted numerous philosophers and physicists. So did Alexandria, Egypt, and Crotona, Italy.

They joyously wallowed in words, sparring in debate, reading their works aloud at the Olympics and other Great Games, polishing their theories to a high gloss, and defending them in print. Their book titles (sometimes all we have left of those works) show the zeal of these early inquirers: they wrote on smells, comets, volcanic eruptions, fainting, old age, giant bones, animals that gore, litigation, the gods, each other. (Sad to say, the lottery of time bypassed Thales and others, leaving nothing of their work except excerpts and mentions in others' books.)

From the intellectual framework constructed by these men (and a surprising number of bold women, from Aglaonice to Arete, Hipparchia to Hypatia) came vocabulary and concepts now indispensable to us: *logic, hypothesis, enigma, idea, criterion, symbol, stoicism, cynicism, skeptic, platonic, utopia.*

In the second century B.C., after Greeks had long been the pinnacle of scientific thought, the Roman powerhouse arrived. Ever practical, they followed the Greek model, then zigzagged in succeeding centuries to exploit technology, ignoring most efforts at pure science research. Only a few protested the stagnation, one being Pliny the Elder, a military careerist and encyclopedist turned science buff, who would later expire from a too-close encounter with an erupting Mount Vesuvius. He denounced the anti-intellectualism of his day, saying, "In spite of official patronage, no addition whatsoever is being made to knowledge by means of original research, and in fact even the discoveries of our predecessors are not being thoroughly studied."

At first glance, it seems clear to us why these ancient societies didn't make more scientific progress. Their dependence on an enslaved workforce, for example. Slave labor undoubtedly did undercut demand for more efficient machines and better use of draft animals. In addition, there were a surprising number of war-free years during Roman-dominated centuries. Peace created an urgent need to keep men in those huge standing armies busy with road building and other long-term projects.

Often we overlook a key fact: Most ancient literary output was written by aristocrats, who emphasized the rigid barriers between pure science and its applications. These armchair erudites would have snorted at the idea of physical proofs. Testing hypotheses, setting up experiments, getting reproducible results? Please. Such hands-on activities were for *technitai*, the technicians—a disparaging label given to inventors as well as blacksmiths.

Such writers would point to the misguided efforts of Heron of Alexandria, whose da Vinci-like ingenuity created a working prototype of the steam engine. Like most of his inventions, it went nowhere. Instead, Heron's dreamworks largely served to amuse as toys of the wealthy, or amaze with whiz-bang special effects at the temples of one god or another.

Theoretician Archimedes, the Einstein of his age, became another example. Although obliged to spend much of his time developing weaponry, he also pioneered mathematical physics while perfecting the major mechanical underpinnings of technology. Three centuries after his death, the Greek writer Plutarch asserted that Archimedes had so despised his practical inventions that he refused to write down how they were made.

Maybe it was Plutarch who had the attitude problem. He was one of a host of historians and scientists from his time right down to ours, who admired long-ago scientific theory yet dismissed its execution. As you'll see in this book, there is growing evidence that early imagineers not only theorized but put their theories into sophisticated practice.

Take the conundrum of the Antikythera Mechanism, only now understood to be a complex calculating machine/analog computer. Its very complexity implies that it could not have been the first device of its kind.

But several factors did blight the full flowering of ancient science. Exhibit A? The opposing core values held by Greeks and Romans. The real estate we call "Greece" today was once a mix of land and sea, dotted with small, wary entities called city-states. They fought, formed alliances, and backstabbed, never forming a single nation despite their shared tongue and cultural beliefs.

When city-states ran low on resources, they'd send colonizers abroad. These far-flung bits of Greekness took hold in Gaul, along southern Italy's coast, on Sicily, around the Iberian peninsula. Dozens dotted the Black Sea shores and flourished in Asia Minor and along the North African coast. Thales' city-state of Miletus alone gave birth to ninety colonies.

That fierce go-it-alone spirit made the Greeks vulnerable—first to Alexander the Great, then to his rapacious successors. And finally, to Rome, who in 196 B.C. conquered them, at times enslaving Greek populations. The better educated losers became the teachers, doctors, architects, and artisans of the winners. Over time, countless thousands won their freedom. Respect, however, was harder to win. Elite Romans educated their sons in Athens but remained contemptuous of Greek intellectual power. Emperors like Hadrian, who appreciated Greek culture, were ridiculed for not having enough Roman gravitas. Behind his back he was called "Greekling," a tag with a tinge of slavishness that implied loser, lightweight, too artsy by half. Small wonder, then, that few young Romans cared to seek careers in the early sciences.

A second somber thread ran through Greco-Roman society, one that severely undermined attempts to find rational answers to scientific problems. We'd call this tangled skein of beliefs superstition.

The majority of ordinary folks around the Mediterranean paid little attention to scientific bulletins or philosophical pronouncements. Instead, the Greek- or Latin-speaking masses alternately marveled at the world or cursed it, shrugging off the idea of seeking to understand. Weather, the fate of a newborn, a harvest, a journey; illness, slow death, or miraculous recovery—all events were in the laps of the gods.

We're your destiny, baby—and can snap your thread any time we like.

Nearly everyone also believed in weaver-goddesses called the Fates, picturing them as three crones spinning thread, measuring it, cutting it to end a human life. In Latin, the root of our word *destiny* meant "that which is woven or bound together with threads." Today when we declare that an event is "bound to happen," we're actually invoking the ancient binding power of destiny.

This universal belief in the implacability of fate meant that the game of life was fixed from the beginning. The only sensible action in the here and now was to placate the powers that be. An easy task, since the ancient world teemed with gods and goddesses, from the dysfunctional Olympians and their extramarital antics to state-endorsed patron deities, demigods, and legendary heroes who could be hit up for favors.

Deity worship eventually extended to regular humans, beginning with a dead emperor or two, followed by living ones. The second-century reign of Emperor Hadrian saw a first: an empire-wide chain of temples to the newly deified Antinoos, the emperor's gorgeous curly-headed boyfriend who'd gone for an unsuccessful swim in the Nile.

As they went about their everyday lives, ordinary people coped by carrying out magical actions. They sought help with decisions through divination of all sorts. For supernatural aid on small stuff, quick answers could be had by mirror gazing or casting lots. For major matters, such as marriage or illness, it made sense to pay an augur for a bird flight interpretation or sacrifice a lamb to read its shining entrails.

Adults from all walks of life firmly believed in the evil eye, in prophetic dreams, in love charms and curse tablets, in malignant ghosts that needed cosseting. The Greeks in particular spent a remarkable amount of time warding off bogeymen and -women, outwitting vampires, combating malign glances, and pacifying sinister forces at crossroads. And even more time, telling horror stories about those who failed to do so.

The pseudoscience bestseller? Astrology, which ultimately garnered the most widespread following, starting in the last century B.C. with an easy-to-follow horoscope based on the zodiac sign at the time of birth.

This bargain bin of options produced a supermarket of superstitious wares, a giant Walmartian chain of magical answers that operated 24/7 around the Mediterranean basin. By comparison, the philosophical seekers, the elite academies of Plato and Aristotle, the mathematical trailblazers from Euclid to Hypatia, and the generations of science researchers at the Great Library and Museum of Alexandria were low-traffic, mom-and-pop stores of precious, often ignored knowledge.

With our historical hindsight, we're amused by the naiveté of the Greeks and Romans, seeing the flaws in their mix of logic and supernatural beliefs. Time for a reality check: magical thinking continues to have a firm grip on cultures around the world, American society included.

Studies during the past decade at Harvard, Cornell, and Princeton have revealed an unsettling amount of irrational thought among college-degreed young adults. Psychologists examined the large number of rituals that people habitually perform throughout their waking hours. They found that such rituals—from wearing lucky clothing to avoiding certain actions—buoyed participants' spirits, soothed fears, and warded off mental distress at threatening situations. They also found that even the well-educated brains of science majors loved to make snap judgments, being quick to link coincidental events as cause and effect—for example, "The day after I began praying for her recovery, she emerged from the coma." Or conversely: "In my argument with Grandma, I called her an old bat, and that week she fell and broke her hip."

Although we'd prefer to think otherwise, we're deeply tied into that ole black magic when it comes to common superstitions. In the United States and the United Kingdom, Friday the thirteenth still causes consternation—

and huge work absenteeism. Avoidance of bad-luck traditions is routine for millions of Americans, such as walking under ladders, opening an umbrella in the house, breaking a mirror, or leaving a hat on the bed, as is the more recent belief that failure to pass along chain e-mails will bring ill fortune.

Good-luck rituals have equal staying power: touching or knocking on wood, throwing salt over the shoulder, or picking lottery numbers that have personal significance. Other studies confirm that we twenty-first-century sophisticates tend to believe that our wishes for the success—or failure—of something can influence the outcome, whether it's a football game or an election.

Like millions of men and women in ancient times, we cling to horoscopes and astrology for financial matters, job advice, and affairs of the heart. What the stars foretell is firmly imbedded in our consciousness—and our wallets. A majority of all U.S. newspapers (and untold galaxies of Web sites) carry daily horoscopes, dozens of magazines devote themselves entirely to astrology, and the American Federation of Astrologers boasts 3,500 professional members. On both sides of the Atlantic, astrologers with real star power reach millions of readers and reap astronomical sums for their prognostications.

Despite the magical thinking that infected millions in the Greco-Roman cultures and that still pervades our modern societies, scientific reason, commonsense beliefs, and logical actions did triumph at times—and in our times as well. We still have much to learn from ancient successes, and much to heed from their failures and excesses.

Take, for example, this clear-eyed assessment from the Greek historian Herodotus. Twenty-five centuries ago, he observed the ferocious, highly mobile Scythian people around the Black Sea and a shudder went through the jovial historian. His prescient words about the hordes that would later overrun his society could apply to our own world today: "The Scythians were more clever than any other people in making the most important discovery we know of concerning human affairs, though I do not admire them in other respects. They have discovered how to prevent any attacker from escaping them and how to make it impossible for anyone to overtake them against their will. For instead of establishing towns or walls, they are all mounted archers who carry their homes with them and derive their sustenance not from cultivated fields but from their herds. Since they make their homes on carts, how could they not be invincible or impossible even to engage in battle?"

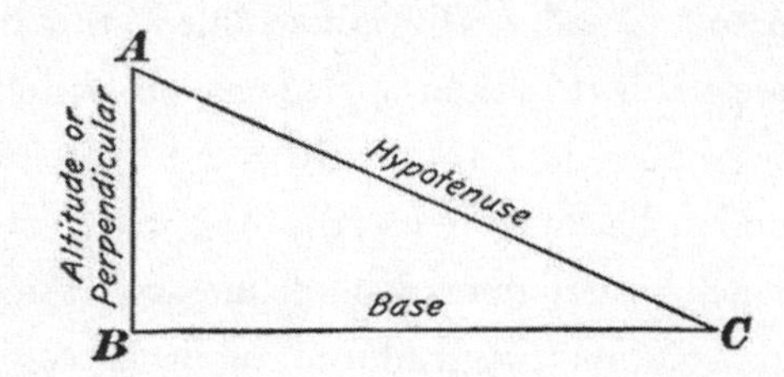

Ancient instruction manual: to make gears, first find a hypotenuse.

For half a century, I've been besotted by the Greeks and Romans of long ago. Although it took years to come to fruition, my own destiny was to learn more by living among their descendants in Mediterranean lands, studying the ruins and remnants of their ancient civilizations, researching and later writing about them. While digging into material for this book on their cultures, I got caught up in the often invisible dance between scientific theory and practice, the push-pull between Greco-Roman thought and seat-of-the-pants advances in applied technology.

It got me thinking about the hypotenuse, a concept I memorized in high school and then promptly deleted from memory. Pythagoras is credited with the mathematical proof that the square of a hypotenuse of a right triangle equals the sum of the squares of the two other sides. Neat. But so what?

Gradually I saw how Pythagoras' theorem paved the way for others to expand on that knowledge—or use it for practical matters. For example, with this shiny new tool, a mathematician named Archytas could tackle a pint-sized but intractable mechanical problem. In order to make gears, Greeks needed metal screws. In order to obtain metal screws, they needed a precise pattern for cutting a spiral into each screw. How could such precision be done by hand?

With the theorem in mind, Archytas assembled a thin bronze rod and a pliable sheet of copper. He cut a right-angled triangle from the copper. After placing one of the smaller sides against the rod, he carefully wrapped the copper sheet around it. The hypotenuse, the longest side, now made a spiral around the rod, which he used to mark guidelines for cutting a spiral into it. Eureka! An elegant solution that produced the first precision metal screw. From that small artifact, the Greeks went on to master gears—and from there to such inventions as the intricate Antikythera Mechanism, built between 140 and 100 B.C. Its function? To make accurate predictions of the movements of the heavenly bodies as the Greeks understood them. And that immense payoff all began with a hypotenuse.

Science writing, ancient or contemporary, can be daunting. And dull. The nineteenth- and twentieth-century classicists and historians who wrote about the Greco-Roman world often dwelled on the boundless, contradictory theories of the ancients. Amid those learned accounts of trials, errors, and blind alleys, the flesh-and-blood human beings, the ones who sweated and swore over those achievements, often got shuffled onto a siding.

To limn a culture, history should attempt to capture the human stories, to squeeze the meaning from cascades of events instead of one thing in isolation. Rather than focus on the elite figures, I tend to stare at the messy, muddy, exhilarating, exasperating mass of humanity of long ago. As a result, within these pages you won't find separate entries on much-written-about aristocrats such as Plato and Aristotle; instead you'll get a sense of their characters—what they desired, what they loathed, what drove them crazy—through their interactions with other seekers.

To give you the true flavor of those times, I've salted this whole brew with little-known facts, Herodotean digressions, and absurdities, once firmly believed yet so outrageous that I could not have made them up. It's an iconoclastic mix, yeasty with names and deeds and beliefs you won't have heard much about. Meteorite worship; bean taboos; bizarre beliefs about women and their powers over hydrocarbons; it's all here.

To the Greeks and Romans, the veil between the world of the living and the dead, seen and unseen, was silk-thin. So here you'll also find a cornucopia of the stories they told and retold about ghosts, monsters, and other marvels.

This book also contains a generous number of flash-forwards, allowing us to appreciate their mistakes, from the overuse of lead to deforestation. These contrasts and comparisons give us a chance to see modern science in the ultimate historical perspective.

In countless ways, the Greeks and Romans were just like us, their blend of two-thousand-year-old science and superstitious beliefs showing remarkable parallels with our own. Although their word for *fascination* (the root of ours) had more to do with averting the evil eye—and using the image of a flying phallus to do it—we can't help but be fascinated in the revelatory sense by the Greek and Roman presence, and their amazing aliveness to possibility.

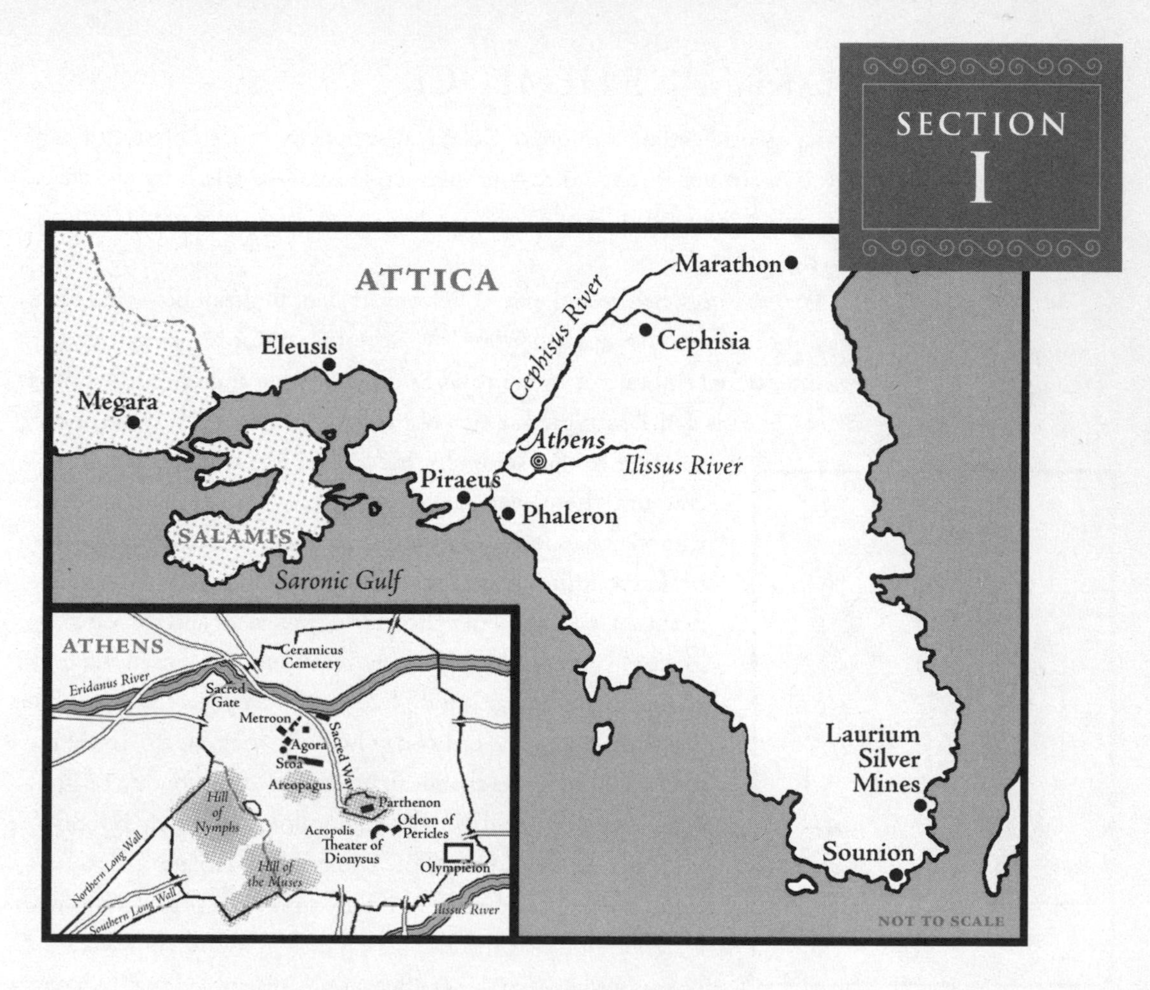

Athens & Attica

MAKE MINE HEMLOCK

Today's world could use more of Socrates' "simplicity is best" lifestyle. Each day as he strolled through the agora (marketplace) of Athens, he would gaze, amazed, at the multitude of wares on sale and say, "How very many things I can do without!"

For centuries, he's been the most beloved ancient philosopher—but that may not have been so in his lifetime. We admire him for his simplicity, for his intellectual embrace of women and humble citizens and slaves. A stonemason like his dad, he sculpted several of the female figures that stood at the entrance to the Acropolis. He looked like what he was: a veteran of Athens' wars, a guy who didn't care about externals, who went barefoot and wore a shabby garment and asked questions about life's meaning. Socrates was married twice, had a couple of kids, loved to dance. He drank water and wine, but he looked like a beer drinker. To us, his round, pug-nosed face and chubby body appear kindly, like an average Joe.

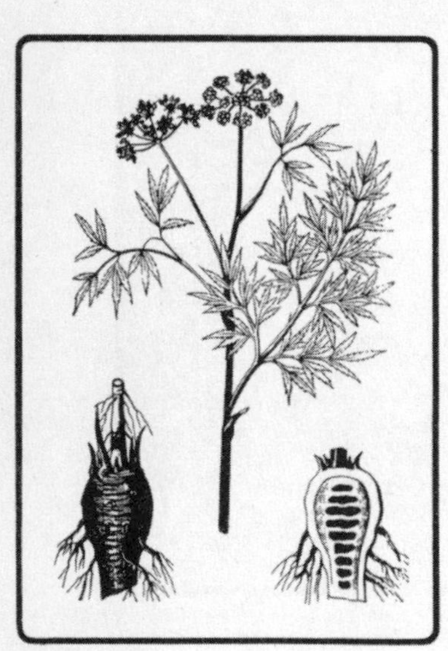

Hemlock cocktails guaranteed painless suicide—at least according to onlookers.

Most Athenians of his day held different views. To them, physical beauty was the ideal. It wasn't merely good to be gorgeous: beauty itself *was* virtue, goodness. Homely Socrates became a figure of fun in the comedies of Aristophanes. His looks evoked half-horrified fascination and scorn. Countless images of him were made in his day and afterward, some because he was well known, others because he resembled that hideous old Greek satyr Marsyas. (For similar reasons, Greeks and Romans produced thousands of statues of dwarves, hunchbacks, satyrs, battered boxers, and other grotesques they considered deformed. Besides amusing the viewer, deformity was apotropaic, meaning it was thought to ward off evil.)

Greek aristocrats resented the upstart, questioning nature of this humble citizen's teachings. They got even more aggrieved when Socrates began attracting aristocratic followers: Plato and Xenophon, among others. They grew positively livid when the circle grew to include a boy prostitute, a cobbler, the sausage maker's son, and infamous women like Aspasia, the longtime companion of Pericles, Athens' military and civic leader.

The Q-and-A of Socratic dialogue that this wise and merry man made his trademark is well-plowed ground. Less well known is Socrates' attachment to democracy—and its defense. A lifelong Athenian who kept himself in shape at the gymnasium, he never traveled to exotic lands to study with sages, an activity considered obligatory for students of philosophy. His only out-of-town journeys were to the battlefield. He fought at Amphipolis, at Potidaea, and at Delium, where he stepped in and saved the life of Xenophon, one of his longtime disciples.

He taught that the only evil is ignorance; the only good, knowledge. That and his self-awareness made the Pythia, the prophesying figure at the oracle of Delphi, say of him, "Of all men living, Socrates is most wise."

Oh boy, that did it. Peer and political jealousy spread like chickweed. The demagogues of Athens accused him of as many crimes as they could think up. Their affidavit is still kicking around, declaring in part: "Socrates is guilty of refusing to recognize the gods recognized by the state, and of introducing other new divinities. He is also guilty of corrupting the youth."

The defense of Socrates got off to a wretched start. His pupil Plato started mewling about how he was among the youngest men ever to address the judges and jury, whereupon the judges yelled, "Sit down!" Soon after that, a majority of jurors voted Socrates guilty. To begin with, they didn't have the death penalty in mind. As the judges mulled over what sort of fine to assess him, Socrates first offered to pay twenty-five drachmas, then said, "Considering my services, you should maintain me at public expense."

This piquant poke in the eye was enough to enrage everyone: instead of exile and/or a fine, Socrates got a sentence of death and was put into prison. A few days later, after saying farewell to his wife, and in the company of friends, he drank hemlock brew. In Plato's account, he walked around until his legs were heavy, then lay down until the drug numbed his body and reached his heart. In a couple of hours, he was dead.

Despite the eyewitness accounts, the world continues to hash over the issue. Did he really drink hemlock? Was it as easy a death as Plato made it look?

What Socrates drank was a potion containing *Conium maculatum*, poison hemlock. Its active ingredient was coniine, an alkaloid that paralyzes the muscles (its action is similar to that of curare, which acts as a muscle relaxant in modern surgery). Conium was used back then to alleviate joint pain, but the difference between a helpful dose and a lethal one was slight. Nevertheless, the plant and its powers were well known. Conium was ground up and sometimes mixed with opium to make a lethal dose.

A later account of another hemlock death from Nicander, an army doctor in the second century B.C., appeared to contradict Plato's story. Nicander wrote that hemlock "makes the eyes roll, a terrible choking blocks the lower throat, the victim draws breath like one swooning, and his spirit beholds Hades."

Research, however, from the author of *The Trial and Execution of Socrates* points out that cicuta, the water hemlock that Nicander described, is another plant whose name is often conflated with that of conium. It can produce tetanus-like seizures, most definitely a nasty way to die. That was not the case at the execution of Socrates.

The Greeks of old used to say, "Do not call a man happy until you know the manner of his death." Perhaps Socrates could have gotten off with a lighter penalty. But he preferred to make a point, regardless of cost. The drama of his death and the splendid writings of his disciples gave Socrates another sort of life, an immortality that few are granted. He'd probably wink and agree.

CALENDAR WARS

Greeks, those rugged individualists, never saw eye to eye when it came to distinguishing one year from another. Or even one month from another. Their calendars varied wildly from one city-state to the next.

Athens, for example, began its year at the summer solstice with a thirty-day month it called Hecatombaeon, then alternated months of twenty-nine and thirty days. Just up the road in Delphi, locals brought in their new year in September with the month of Bucatius. The Spartans started their year in October with Herasius. And so it went. Months had different names and lengths, years had different start dates.

This freakish individualism extended beyond the calendar. In ancient times, there was no such animal as Greece. Most men and women lived in towns or smallish cities called city-states—*poleis* in Greek (sing. *polis*, from which words such as politics and police derive). Whether located on an island in the blue waters of the Mediterranean or in the landlocked cornfields of Sparta, each city-state was an island

A decent calendar? Don't ask me—I've got bigger problems, in case you hadn't noticed.

of independence. When a *polis* grew to a cumbersome size, it burst like a milkweed pod, sending seeds for new colonies west to Sicily and Spain, south to populate North Africa, east to Syria and Asia Minor, northeast to circle the Black Sea. In time, the colonies themselves sent out settlers. Over centuries, the *poleis* rose and fell, allied and fought, coupled and uncoupled. Small wonder that they never adopted a mutual system of marking days, weeks, and years.

Athens, the *polis* with the largest population, limped along with a lunar calendar of 360 days, sticking in a thirteenth month every second year—a slipshod arrangement that left the year too long or too short. About 594 B.C., the statesman and lawmaker Solon introduced a Goldilocks model that he thought was just right. It still had alternating months of twenty-nine and thirty days, but three times in each eight-year period an extra month was inserted to soak up the excess. The Spartan city-state, largest in acreage, covered most of the southern chunk of the Greek mainland called the Peloponnesus. Its lunar year differed substantially from the year of its Athenian rivals.

Given this temporal haziness, the Greeks grappled with other means to keep track of long periods. Some city-states marked the passage of years by naming them after a given political or religious leader during his or her term of office. Annually in Athens, nine names were drawn out of the hat for the office of archon, or chief magistrate; one lucky guy became the eponymous person after which that year would be called. If plague or political disturbances interrupted the orderly selection of new officials, those periods got labeled "year without an archon," or *anarchon*. Our modern usage of anarchy takes on a different flavor when these woeful underpinnings are recognized.

At a certain point, however, Greek city-states sullenly agreed on a universal chronology, one that was honored by even the most bellicose: the celebration of the Olympic Games every four years. The traditional beginning date for the Olympics was 776 B.C.: from the get-go, a meticulous victors' list was kept. Thanks to the Greek go-it-alone attitude, however, it didn't cross anyone's mind to use the Games for tracking years until the fourth century B.C. Once that transpired, folks had an easy, nonlocal way to fix points in time, as in "that happened in year three of the 102nd Olympiad."

The Greeks clearly had calendar issues, but no worse than the mess the Romans got themselves into. Despite their engineering skills and pragmatic ways, they waffled greatly when it came to time. They too struggled with lunar versus solar calendars, start dates for the new year, and fall celebrations that ended up in the spring.

The Romans brought more calendar grief upon themselves with a barbarously complex day-naming scheme. Days had dual sets of names; the first were astrologically tied to the Big Seven stars and planets (sun, moon, Mars, Mercury, Jupiter, Venus, Saturn), but apparently that seemed too facile. On the official Roman calendar, only three days of lunar significance were identified: the first day of the month, called *kalends;* the fifth or seventh of the month, called *nones;* and the thirteenth or fifteenth of the month, called *ides*. Instead of saying March 10, Romans would call it "day five before the ides." Things got worse in print. In order to copy down the date we would call January 15, a Roman would be obliged to write "A.D.XVIII.Kal.Feb," or "eighteen days before the kalends of February."

Whether they lived on an island, in Athens, or in rural Italy, farmers, mariners, and ordinary folks still kept track of Father Time in time-honored ways: by moonrise and sunsets, the passage of the seasons, the solstices, the equinoxes; by festivals and market days; and by days of good luck or ill omen, as announced by city officials or religious elders.

They also leaned on the *parapegmata*, a body of oral wisdom (later in more permanent format) that tracked the larger cycles of weather, seasonal winds, star movements, animal behaviors, and day lengths so that crops could be sown at the right times and sea traffic could move during the safest periods. The Roman author and historian Polybius left an interesting example of such usage. In 255 B.C., a marine disaster he called the biggest ever involved the loss of 284 ships. As Polybius pointed out, the expedition's military commanders had ignored the *parapegma* advisory that warned ships not to sail between the rising of the star Orion and the rising of Sirius, the Dog Star.

Frankly, both the Greek and Roman systems sound like major headaches to modern ears, but it's all about what you grow up with; perhaps our complex mix of printed calendars and school years, fiscal periods and charge-card cycles, red-eyed LEDs and robotic voices from digital devices would appear the same way to them.

What's more, their timekeeping also had a very human dimension that we've lost: a vast army of town criers and priests, who announced the times of day and the important dates in each month in voices anyone could hear.

THE SCIENCE OF SEEING

Greeks of long ago (and not so long ago) used to wake up to the most brilliant of skies, to azure horizons where one could see forever. Perhaps the clarity of that vision led them to their aesthetic achievements—the science of seeing, that very Greek sense of the beautiful.

Case in point: the Parthenon and its Doric sister temples in other parts of the Greek world. Despite its battered beauty, its missing pieces, the

Gorgeous visual trickery at the Parthenon, plus ample storage for booty, bank deposits, and thank offerings.

Parthenon still stands like a haughty queen on her throne above the sharp vertical face of the Athens Acropolis.

Begun in 447 B.C., spearheaded by the charismatic leader Pericles, built by the daring architects Ictinus and Callicrates, dressed by the sculptor Pheidias and a crew of Athens' finest artisans, the Parthenon raced to completion in fourteen years—an eyeblink of time. Now denuded, it once housed a colossal gold and ivory statue of Athena, guardian of the city. After Pheidias sheathed the statue in gold, Pericles (wise to the ways of the litigious Athenians) suggested that he make the gold plates removable—all 44 talents, or 2,532 pounds, of them. Good advice. In 433 B.C., Pheidias was indeed accused of embezzlement but declared innocent when officials found the plates to be intact.

The Athena image occupied just one room. As with similar buildings, the Parthenon also served as a bank in ancient times. Thus it had storage areas filled with chests, crammed with coinage. Moreover, it functioned as a storehouse of treasures and thank offerings. A partial inventory: six thrones, ten couches, seventy shields, Persian daggers, gold and silver bowls, gem-studded figurines, war booty, and city heirlooms. Archaeologists have also identified places where metal security grilles were once installed, confirming that parts of the Parthenon remained inaccessible to the public.

Junk to jewelry, it's all vanished now. But the greatest treasures once held by the Parthenon were its sculptural friezes, 160 feet of them, circling the building. They depicted a still-enigmatic procession, whose muscled animals and human figures leap from the stone, leaning toward the viewer.

The Parthenon didn't really need gifts or accessories, however; the building itself was created to exalt the viewer, to completely satisfy the eye. Modern-day restorers continue to be surprised by the architectural secrets and engineering technology it reveals.

For instance, those columns may look straight and upright, but they're not—and never were. Each is slimmer at top and bottom, swelling subtly in the middle like a cigar. The outer columns lean inward, giving the visual effect of stability and aesthetic rightness. The floor has a gentle curve to it as well. This technique creates architectural tension, called *entasis* by the Greeks.

The Greeks evidently relished challenges. Not only did they build with marble, using the simplest of tools and plans, they aligned the columns without using any mortar between the drums. To achieve this, the artisans shaved away the areas of contact by making each drum concave. When assembled vertically, the stacked drums touched each other only on a three-inch rim of their outer edges.

The original stonemasons also spent endless hours carving long vertical grooves into each main column—then lavished more time polishing and then texturing each exposed surface with chisel marks in orderly rows. It's now believed that they worked twice as fast as modern stonemasons, in part because they possessed sharper, harder chisels and axes. None of their hand tools has turned up yet, but the term *by hand* takes on a whole new level of meaning after close study of the Parthenon.

The current generation of restorers have had a dauntingly slow go, largely because earlier efforts almost destroyed the Parthenon. The original builders had fastened the marble columns and building blocks together with invisible iron clamps, afterward sealing them from corrosion with molten lead. In 1898, the archaeologist charged with restoration installed new clamps of iron but failed to seal them. They rusted and swelled, cracking much of the marble.

When this historically accurate restoration of the Parthenon is complete, Athena's treasure-house will conquer new generations of admirers. In the meantime, those who long to see more of what the long-ago Greeks built and thrilled to should visit Sicily. There, by great good fortune, a whole array of classic Doric structures still shimmer in the sun, most of which predate the Parthenon. Their monumental columns grace the landscape from Agrigentum's Valley of the Temples to Selinunte and Syracuse.

THE WHOLE NINE CUBITS

As the canny Greek philosopher Protagoras (whose own name meant "first in the marketplace") once said, "Man is the measure of all things." Yes, and the measurer as well. Throughout history, humans have used their body parts (and their physical limitations) to set up standards of size, weight, length, and distance.

Along with those upstart Phoenicians who expanded from their own bit of real estate on the eastern Mediterranean, the early Greeks muscled their way into exploration, which led to imports, exports, and the first trade deficits in history.

Right away, these venturesome traders and explorers realized they needed standards. Athenian traders had olive oil to sell, and glorious pottery. The Greek colonists on the Black Sea had vast amounts of wheat and good wine. These items and countless others required a system of consistent, honest measurement between buyers and sellers.

The traders faced another stiff challenge. There was no PayPal, little credit, few middlemen to serve as guarantors. People struggled to come up with mediums of exchange. They experimented with the island of Cyprus' big copper ingots in the shape of oxhides, but early adopters groused about the awkward shape, the storage issue. "And they turn all green after a while" was a common complaint.

Near Athens, someone produced another innovation: iron spits called obols, which proved nearly as cumbersome. They tried bundling them in groups of six, called a drachma. Underfoot, the darn things barked shins and created another storage headache. (The Athenian Greeks would later recycle the words *obol* and *drachma* as names for several of their coins.)

Speaking of things metallic: in the seventh century B.C., a sleek new medium called coinage began cropping up, launched first by the kingdom of Lydia, then adopted by a few Greek city-states. This medium used pocket-sized pieces of silver and bronze as a way to convert product values and pay for them. One drawback: no one had pockets yet. The second, more serious hitch? Coinage wasn't necessarily accepted anyplace but at home.

Given the prickly city-state system, where each cluster of Greekness was by turns the ally, enemy, or competitor of the others, that development wasn't surprising. The only exception? The charming coinage of Athens, called "owls" for the grave totemic bird on each silver circle. They came to be accepted almost everywhere.

Besides the rancorous issue of a medium of exchange, the need for standards for weighing and measuring became just as pressing. The Greeks devised weights standards, using squares of lead marked with symbols such as turtles or dolphins to denote various quantities. In the agora itself, officials began to oversee commercial activities to make sure that buyers weren't cheated, checking scales for correct weight, and even spot-checking goods for quality.

The only fly in the buy-sell ointment: each city-state's mania for establishing its own set of weights and measures. Soon there were the Argos, the Aegina, and the Attic (Athens' own) systems. The Attic mina weighed about one pound; the Aegina version weighed two thirds more. In short, a nightmare for importers and exporters. No wonder the Greeks got to be such whizzes at math.

Although traders groaned at the taxes, they were probably delighted when Romans finally imposed their system of a *libra* divided into twelve parts called *unciae*. (Our words *ounce* and *inch* both come from *unciae*.)

Measurements of length encountered similar quarrels. Depending on city-state, the Greeks had half a dozen versions of the foot and the *stade*, the latter referring to the length of the official footrace at Olympia, site of the Olympic Games. As a structure, the stadium didn't exist yet, being a later development. Depending on where you were in the Greek world, the *stade* varied from 583 to 630 feet—if you happened to be the runner, that was a dire difference.

And the cubit? Patterned after the length from a man's elbow to his outstretched fingertips, its variations ran the gamut. The early Greeks began with a 14-inch cubit, gradually creeping up to 18-plus inches for the Olympic Games' standard cubit. Romans went with 17.4 inches. Noah and the Old Testament crowd thought of the cubit as 17.5 inches, whereas by the first century A.D., Jews asserted that their cubit was 20.6 inches.

Even the ubiquitous amphora, used by everyone around the Mediterranean to hold liquids and grains, had three different standards. On Crete, an amphora held 20 to 24 liters; the standard Roman amphora was 26 liters. By the time you got involved with Greek amphora, you were looking at a whopper of 39 liters. A gracefully shaped workhorse, the earthenware amphora was sturdy enough to travel by ship and light enough to be moved by one husky stevedore. Ships were actually designed around amphorae loads, which determined the shape of the hold, since amphorae were packed in a herringbone pattern, nested in moss or pine needles for the voyage.

In the city of Rome, a large artificial hill called Mount Testaccio still stands, its grassed-over contents a testament to the incredible amount of commercial activity in the ancient world. The hill is constructed entirely of pottery shards from millions of amphorae. During Rome's centuries of vitality, amphorae filled with oil or wine or grain arrived at the always-hungry city. Once used, each amphora was carefully broken into shards and stacked. More pragmatic than environmentally minded, the Romans had discovered that it took less work to manufacture a new one than to cleanse an old one—especially since a decent detergent that could cut through rancid oil residues and wine lees had yet to be invented.

Fellow cynics, I've written a hot new diatribe. Let's hit the streets!

FAR-FROM-CYNICAL SOUL MATES

The city of Athens never quite recovered from the antics of Crates and Hipparchia, an odd couple whose alliance was philosophical, physical, and fun, by the sound of it.

Crates was born in Thebes, Hipparchia in Maroneia on the Black Sea. Both grew up in aristocratic households where big names, such as father-son world conquerers Philip II and Alexander the Great, dropped in for casual visits.

Once educated, Crates drifted down to Athens, an idealist hungry for challenges. He connected with Diogenes, who followed the school of thought called cynicism. Back then, cynicism carried more than its modern cargo of contemptuous disbelief in human goodness. Feisty old Diogenes practiced a grubbily austere way of life, criticizing consumerism and material pleasures.

"To become a true Cynic, you'll need to throw your inheritance into the sea," Diogenes warned. Instead, Crates turned over his assets to a banker, saying if his sons proved to be ordinary men, they'd inherit. If they became philosophers, the banker would distribute the wealth among deserving people. As Crates saw it, his sons would want for nothing if they took to philosophy. A far-reaching idea, especially since Crates didn't even have a girlfriend yet, much less a kid on the way.

Patterning himself after his teacher, Crates adopted a diet of water and lentils. His uniform of choice was a ragged cloak over his swarthy, ungainly body.

At length Metrocles, the kid brother of Hipparchia, came to Athens to study philosophy at the Lyceum, the school founded by Aristotle. Before long, Metrocles ran into the perennial dilemma of students: even with an allowance, he couldn't make ends meet. As a Lyceum undergrad, he was expected to have servants and fancy furniture, besides paying his share of all those wine-soaked philosophical dinners.

To economize, Metrocles started eating the cheapest thing going: beans. While giving a class speech one fateful day, his diet noisily backfired. Disgraced, Metrocles ran from the classroom, intent on starving himself to death. Starvation, however, turned out to be surprisingly time-consuming.

One day as Metrocles brooded on his wool mattress, he heard someone knock. There stood a man in a disreputable cloak: Crates. He gently poked fun at the boy, then demonstrated his own skills at flatulence. Instant bonding. Metrocles quit the Lyceum and became one of Crates' students, raving about him to his family, who urged them to visit.

When Hipparchia laid eyes on this homely fellow twenty years her senior, she was struck by the thunderbolt, that Greek helplessness in the face of passion. Her more suitable suitors suddenly meant nothing. She wanted Crates, a man whose only assets were kindness and a rollicking sense of humor. She threatened to kill herself if her parents would not let her marry him. (Like her brother, she had melodramatic tendencies.)

Resorting to shock tactics, Crates stripped off his cloak, saying, "This is my only possession. Come with me, and you'll share my hardships."

This gently bred girl rose to the dare. They returned to Athens to wed on the steps of the stoa or Painted Porch, the favored gathering place for philosophers. Hipparchia may have gotten her first taste of cynical disdain for convention at her own wedding—rumored to have been consummated in public!

Athenians had nicknamed Crates "the door opener" because he was welcome in every home in the city. Like her new husband, Hipparchia adopted cynicism's more compassionate tenets. She too became a street philanthropist, a solver of everyday problems, consoling the bereaved, dispensing practical remedies to the sick.

She and Crates got a bang out of writing diatribes—the sharp, funny op-ed pieces that cynics used to get the word out. She gained further notoriety for her habit of going everywhere with her husband, even to dinner parties. Respectable women weren't welcome at all-male symposia. Flute girls and sex workers, yes. But Athenian matrons? Not on your baklava.

At one party, Hipparchia had a run-in with a philosopher named Theodorus who tried to rip off her cloak—a contemptuous gesture meant to show others she was a woman for hire. Calmly outmaneuvering this bore, Hipparchia engaged him in philosophical swordplay. She said, "Do you agree that any action that wouldn't be wrong if done by Theodorus wouldn't be wrong if done by Hipparchia? Now, if Theodorus does no wrong by striking himself, then neither does Hipparchia when she strikes Theodorus." She followed up her sophism with a healthy slap.

An angry Theodorus sputtered,"Who is this woman? Why isn't she home weaving?"

"The name's Hipparchia," she answered."Have I been ill-advised to waste my time on philosophy when I could have spent it working on the loom?"

Not as snappy a rejoinder as might be heard on late-night television, but it killed them back in fourth-century B.C. Athens.

Amid the community organizing, diatribing, and dining, Hipparchia and Crates had two children. Like any multitasking mom, she held their penniless household together with willpower and endless batches of lentils, the mortar of their lives.

When their daughter wanted to wed, Crates and Hipparchia had her live with the boy for a month—the first trial marriage on record. Their son, however, claimed his father's inheritance. Sick of lentils and voluntary poverty, he rejected the social protest philosophy of cynicism.

Their standout disciple, however, persevered. Zeno, a humbly born islander, opposed the status quo chauvinism of the schools of Plato and Aristotle. He made a habit of teaching under the shaded colonnades of the stoa, which is how he became known as a Stoic and his evolving worldview as Stoicism.

Five centuries later, a famed Stoic named Epictetus declared that cynics should never marry or rear children—then gave Hipparchia the highest compliment a man of that era could make. His much-quoted line:"Yes, but Crates took a wife, a circumstance that arose from love and a woman who was another Crates."

Today we might add, "And a kind-hearted man who evidently was another Hipparchia."

No need to bellow—even the folks in the nosebleed seats can hear us.

THE LOST ART OF ENTHUSIASM

Like us, the Athenians were dead keen on special effects. No CGI for them. Unlike us with our passive, massive consumption of film, television, online content, and video games, they got their thrills from live productions.

The Greeks participated in theatrical events and religious festivals with an enthusiasm we lack. The word *enthusiasm* itself comes from ancient Greek, its literal meaning "temporarily possessed by a god." Sounds corny, but that was the emotion they sought. They also craved catharsis or emotional cleansing—the kind you get from watching a good tragedy about a prophecy, a boy, his mom, a blinding, and an unseemly love affair.

What sorts of special effects did their theatrical productions employ? The backdrops of early Greek theater began ultrasimply, by the third century B.C. becoming more elaborate with revolving panels depicting new scenes. Later Roman theater went berserk for visuals and fancy stage sets, elaborately painted to produce striking three-dimensional effects.

Athenians, however, understood the most powerful special effect: sound. From the start, the intimacy and acoustic wizardry of their theater architecture let every unamplified word be heard, from a whisper to a shout.

Another visceral punch came from the Greek chorus, fifteen to fifty men who acted as links between the actors and the audience by singing, dancing, and responding to the actors in ways that punctuated the action onstage. They set the mood for drama: reinforcing tragic climaxes, inserting pregnant pauses, raising goose bumps with haunting song. In comedy, they fed straight lines to the actors and, to emphasize the absurd, wore silly costumes or eye-catching leather phalluses.

The Greek chorus didn't simply trot onstage to do a number. They remained half-circling the stage, a visible counterpoint, an audible soundtrack to the entire performance. The choral leader wore wooden shoes with percussive soles, keeping his chorus in unison to provide the musical heartbeat of the play. Their rhythmic chanting could raise the hair on playgoers' arms.

Because stage backdrops showed exteriors only, playwrights invented a tableau machine to emphasize moments of high tension. Instead of having a murder committed offstage, a platform on rollers came center stage at a key moment. On it, frozen in a dramatic pose, would be the victim and the murderer, weapon in hand.

After the era of Aeschylus, Aristophanes, and other major playwrights of fifth-century B.C. Athens, the postclassical theater developed more special effects. The *bronteion* (from which we get *brontosaurus* or "thunder lizard") simulated thunder, along with a companion device that imitated lightning. Trapdoors came into use. One dandy contraption rendered the sudden appearance of a ghost more frightening; it was called the Stairway to Hell.

Modern drama critics tend to hold their noses at plays that wrap things up too neatly in the third act. To the Greeks, who invented *deus ex machina*, the idea of deities leaving their Olympian neighborhoods to meddle in human affairs or carry out a nick-of-time rescue had a special resonance. They wanted to believe in a higher power, whether that meant an angry god or fate. The appearance of deities forced the human actors to answer to that power—or wrestle aloud with their own consciences.

High-minded motives aside, the Greeks also adored the machinery that allowed a god or goddess to appear. Theater technicians used special cranes, stationed behind the proscenium, to carry an actor from on high down to the stage floor, often a two-story flight. Each crane was powered by men using their weight to revolve a treadwheel cage.

In the comedies of Aristophanes, some of the best comic flights of fancy involved actual "flights" via the crane. In his comedy *Peace*, the farmer protagonist rides off to heaven on the back of an enormous dung beetle, trying to get the gods to stop war. Another play featured the hero Bellerophon riding Pegasus, the flying horse. The tragedy *Medea* had the murderous mother whisked away in a chariot pulled by two dragons.

The only trouble with Athenian theater performances? Unlike the months-long runs on Broadway, they were staged during special festivals such as the City Dionysia, the Lenaea, and the Anthesteria. That left a lot of downtime between live performances.

But the Greeks of the classical age got further thrills and catharsis from religious celebrations. Mystery cults such as Eleusis and women's festivals, including the Thesmophoria, weren't predictable, pious affairs. Their ceremonies employed generous amounts of mind-altering substances and activities, from wine and opiates to trance-inducing smoke and drumming. Plenty of enthusiasm to be had by all.

By Hellenistic times—the centuries before and after the death of Alexander the Great, roughly 350 to 200 B.C.—Greeks were becoming more sophisticated. It took more to astonish them and thus their temples (and Roman ones, following the Greek pattern) began to acquire religious gadgetry. Some had entrances that mysteriously opened and closed; others had fanciful trickery, such as blasts from invisible trumpets.

Within the temple precinct, other special effects awaited. Beautifully crafted machinery of bronze or gold, often with jeweled fittings, delighted visitors with divine figures that danced or performed a series of actions. These ingenious automata featured snakes that hissed and struck, birds that sang, archers that shot arrows. Men such as Heron of Alexandria, one of the most talented imagineers, invented a whole array of religious wonder gadgets, commissioned by a given temple.

Later writers have often speculated why the Greeks and Romans "wasted" their time with trivial uses for their mechanical innovations. Those long-ago folks might look at our age, with millions mindlessly inhaling a nonstop flow of special effects, quick-jump visuals, and amped-up noises, and ask the same question of us.

METEOROLOGY-MAD BOTANIST

After observing shooting stars, the Greeks decided to call them *meteoron*. It was shorter than saying, "You know, those things we saw whizzing through the air last night—what was that all about?" By degrees, the term took on more heft. In a couple of centuries, *meteorology* was understood to signify "a discussion pertaining to celestial phenomena."

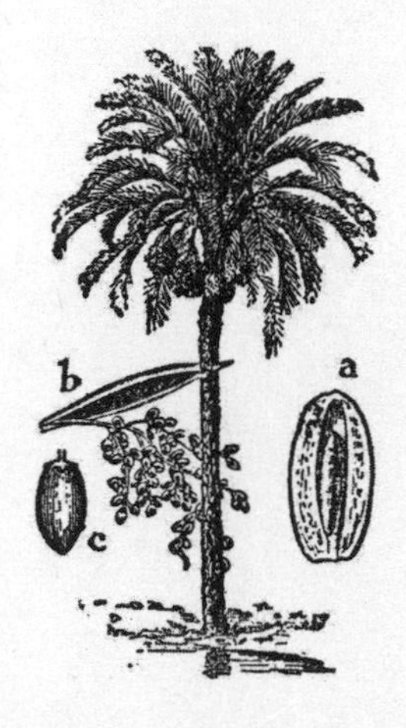

Storm-chaser makes career move to botanist, discovers sex life of the date—and of his own.

They may have gotten their wires crossed when it came to meteors, but at least one Greek weather prognosticator had more firsthand knowledge of climatic conditions than the glibsters nowadays who read teleprompters and slide around electronic maps of cold fronts with such aplomb.

Although he couldn't boast about TV ratings or even a cable show, a likeable natural scientist named Theophrastus knew a tremendous amount about meteorology. To study climatic conditions firsthand, he went out and collected empirical data in the field—which, given the couch-potato habits of many long-ago Greek scientists, was unusual.

At some point, he wrote a book of two hundred useful generalizations, a snappy little number called *Concerning Weather Signs*. It was one of literally hundreds of titles he wrote. (Books written on scrolls were much shorter than books in the current format. They had to be; you were wading through someone's handwriting, after all.)

He and Aristotle, the man he'd studied with most of his life, both authored books on weather. Remember the old saying "Red sky at night, sailors' delight"? That's from Theophrastus—and it's nearly always true. Why? Because red skies, usually seen to the west where the sun sets, occur when air contains very little moisture. From observation, Theophrastus knew that weather systems tended to move from west to east, so red skies meant chances were good for nice weather the following day. What about the "Red sky in morning, sailors take warning" business? True again—as long as the viewer realized that Theo meant a threatening sky, dull red in color, and not a pink dawn, a sign of continued good weather.

Born about 370 B.C. on the island of Lesbos, the son of a humble fuller who cleaned clothes for a living, he was called Tyrtamus as a boy. After he began study with Aristotle at the Lyceum in Athens, he won the nickname Theophrastus or "divine speaker" for his graceful way of talking.

Bright and energetic, this man was described more than once as "ever ready to do a kindness." For those qualities and more, he became the obvious choice to take over the Lyceum when Aristotle died in 323 B.C. The man lived up to his early nickname. At the Lyceum, he drew as many as two thousand pupils to his lectures.

Nevertheless, out in the field, collecting and observing, was where he really enjoyed being. His mentor, Aristotle, had been instrumental in carrying out the first scientific classification of animals. Theophrastus, who believed that animals were capable of reasoning, had differences of opinion with Aristotle. On principle, he became a vegetarian and chose to learn everything there was to know about botany. Most of his books on plants have survived. In them, he classified more than five hundred plants and established the first terminology to describe their structures. This sharp-eyed thinker also explained how plants reproduced sexually—and in one of his books demonstrated how to hand-pollinate the date palm and how to propagate many other species.

Besides becoming the "father of botany" and one of the world's first real ecologists, Theophrastus taught at the Lyceum for thirty-five years. He remained a bachelor, although he did have a stable relationship with a woman named Herpyllis. Their arrangement had a peculiarly Greek twist. A former slave who'd been given her freedom by Aristotle, Herpyllis had been the long-time *hetera*, or sexual companion, of the philosopher. In Aristotle's will, she was bequeathed to good old Theo! Pretty cozy. (He also inherited her children and the property on which all of them lived.) The wills of both men are given in full in the books of Diogenes Laertius, an invaluable surviving source of biographical information on the early scientists and philosophers.

Curious and busy until the end of his eighty-five years, Theophrastus continued to work and teach, conducting his life by his much-quoted maxim "Time is the most valuable thing one can spend."

He spent his well. It does make you shake your head at the vagaries of history, however, when you learn that the important body of work he produced on plant life and climate was pretty much ignored by later Romans, whose idea of an efficacious method of making rain was to throw small straw puppets called *argei* into the Tiber River.

HE GAVE PHYSICS WEIGHT

His given name was Strato, although in his prime teaching and research decades (the latter half of the third century B.C.) many simply called him "The Physicist." A man with a big ego could have gotten puffed up over this, but he wasn't that sort.

In his day, physics meant the study of the underlying unity of the world, including the natural world and its functions—a much broader look across disciplines that are now separate.

Strato became a peripatetic in two senses. He studied at the Lyceum in Athens, the school founded by Aristotle in 335 B.C. The student body got labeled *peripatetics* or "walk-arounds" because that's how the hyperactive

Talk about a sweet gig—I'm surrounded by acres of books and reshelving slaves.

Aristotle had taught—strolling under the cool vine-covered walkways of the Lyceum to escape the blazing sun of Athens. In a traveler's sense, Strato also was peripatetic; like other inquirers into life, during his long career this gifted man commuted between the great cities of Athens, Greece, and Alexandria, Egypt.

Although born in Lampsacus, a Greek city-state in Asia Minor (northwest Turkey), famed as a center for the Priapus cult, fine wines, and not much else, he soon headed for the Big Pomegranate—Athens—because that's where the action was. He studied under Aristotle's successor Theophrastus, who focused more on botany than biology.

Strato had other interests. And big questions. As time went on, he disagreed with a number of the school's ideas. The revered Aristotle had written so voluminously that his teachings—spot-on insights and wrongheaded blunders alike—stifled further exploration by new generations of thinkers.

A fearless, often blunt man who did not mind controversy, Strato didn't think the gods had much of anything to do with the physical world. Instead, he trusted in a marvelous material universe, one whose secrets could best be discovered through experimentation meshed with cogitation. The data he collected and analyzed gave him important insights, one being that all elements had weight and thus were subject to gravitational forces. His original research became a cornerstone for the likes of Archimedes and others in later centuries.

He wrote copiously on such matters as acceleration and vacuums, but little has survived except excerpts quoted by other writers. Here's a bit from his experiments with air, preserved in the writings of inventor-engineer Heron: "Vessels believed to be empty are not really empty, but full of air. Air . . . consists of minute particles of matter for the most part invisible to us. If one pours water into an apparently empty vessel, a volume of air comes out equal to the volume of water poured in. To prove this, make the following experiment. Take an empty vessel. Turn it upside down, keeping it vertical, and plunge it into a dish of water. Even if you depress it until it's completely covered, no water will enter. This proves that air is a material thing, preventing the water from entering . . . Now bore a hole in the bottom of the vessel. Water will then enter at the mouth while the air escapes by the hole . . . proof that air is a bodily substance."

Strato loved to study forces, motion, mechanical phenomena. He discovered, contrary to Aristotle's teachings, that falling objects accelerate as they drop. He wrote, "If you drop a stone from the height of an inch, the impact made on the ground will be small, but if you drop the same object from 100 feet or more, the impact will be a powerful one. The weight of the object itself has not become greater, it does not strike a greater space of ground, and it isn't impelled by a greater external force. It is merely a case of acceleration."

He may also have authored *On Mechanics*, the first book to describe meshed gearwheels, and *On Sound*, both attributed to Aristotle at one time. During his travels, Strato gave serious thought to geology, hypothesizing from the physical evidence he'd seen that both the Black Sea and the Mediterranean had at one time been closed systems—and that in the deep past, continents had been seas, lakes had been bogs.

Before long, Strato fielded a call from one of Alexander the Great's successors, Ptolemy I, who wanted him to tutor his son. Thanks to Strato, Ptolemy II got an excellent grounding in physics and other sciences, along with a real love for learning. Later, when Ptolemy II took over as ruler of Egypt, he founded the world's first research institute around 306 B.C. and made a generous, no-strings-attached offer to Strato. Called the Great Library and Museum of Alexandria (*museum* from "home of the Muses," the nine goddesses of culture and learning), the place would become a dazzling beehive of intellectual pursuit, the Cambridge or Stanford of its day, with the Library of Congress on the side.

Once in Alexandria, Strato worked with a fellow peripatetic, Demetrius of Phalerum, to get the research institute up and running. Besides overseeing the facilities for as many as one hundred scientists to live and study on-site, Strato made sure the place offered real-world studies of flora and fauna. He was the driving force behind the museum's immense botanical garden plus a wildlife refuge that included live chimps, elephants, lions, ostriches, parrots, leopards, cheetahs, Saiga antelopes, a giraffe, a rhino, and a python forty-five feet long.

He also taught there; his work may have influenced students such as

Aristarchus, who became a noted astronomer, and Erasistratus, a pioneer in human physiology.

Back in Athens, Strato's longtime friend and mentor Theophrastus began to fail. Upon his death in 287 B.C., and at his urging, Strato was chosen as his successor, the Lyceum's third director. He ran the school for two decades. In his lean old age, he remained as bright-eyed and blunt as ever, noting in his will, "I leave the school to Lycos, since of the rest some are too old and others too busy. But it would be well if the others would cooperate with him."

WHY ANCIENT FORESTS FELL

Death struggles between nations over natural resources are nothing new. When not fighting over mineral rights, the Greeks went to war over trees—and fought it with trees. To meet the Persians at Salamis in 480 B.C., the Athenians and their allies built a fleet of 350 ships that devoured miles of forest. Not Greek forests, though. Theirs would no longer suffice, so timber had to be imported from southern Italy and regions north of Athens.

During the same war, when Xerxes and the Persians invaded mainland Greece, they deliberately set fire to Greek woods—and continued their destruction on their way home in defeat.

The second Peloponnesian War between Athens and Sparta became a vicious conflict over timber. The Spartans had conquered wooded regions of the Macedonian coast, which the Athenians needed for shipbuilding. We think of Thucydides as a historian, but he was also an Athenian military leader whose mission was to win back valuable timber-covered lands. He failed. As was the custom with generals who goofed up, he was exiled. Only then did Thucydides write his now-famous account of the war.

Perennially desperate for wood, Athens later launched a military expedition to the heavily wooded island of Sicily. Not only did the Athenians grind up many square miles of timber for their own fleet of 134 ships, but they lost every vessel and nearly forty thousand men in that debacle.

History books sometimes put the primary blame for wilderness loss on vegetation-loving goats, sheep, and cattle. But control of the forests represented an ancient tug-of-war, with animal damage causing secondary fallout. The worst part about animal grazing around the Mediterranean? It prevented the regeneration of forests already lost through conflict or overharvesting. (It's still a hot issue.)

Mining activities doomed woodlands as well. Homer and other long-ago poets had sung of Greek heroes wearing the famed bronze armor from Cyprus. At that time, the island wore a thick mantle of green, composed of tall cedars and pines—both sought for shipbuilding. In addition, its earth concealed copper, the key component of bronze. This abundant resource was enthusiastically mined.

Copper and silver mining, the production of charcoal for mining furnaces, and the clearing of wild lands all required the downing of trees. Lots of trees. Oaks and conifers disappeared as the deforestation juggernaut of mainland Greece, of the islands, inexorably rolled on.

In the region of Attica, with the city-state of Athens at its heart, silver mining did grievous harm. In one three-hundred-year period, the mines wiped out 2.5 million acres of native forest as well as using over 1 million tons of charcoal. By 400 B.C., thoughtful Greeks were mourning the devastation of Attic lands, eloquent Plato among them: "What remains is like the skeleton of a body emaciated by disease. All the rich soil has melted away, leaving a country of skin and bone."

Things got no better in Roman times. Vast numbers of primeval woodlands went up in smoke as populations grew and expanded geographically. As much as 90 percent of the wood harvested was burned as fuel—to cook, to heat, and to make charcoal for cooking and heating.

Huge quantities of timber went into construction: roof beams for temples, ship hulls and masts, private housing. Industrial uses chewed up vast amounts, too: making pottery, smelting metals, producing vehicles, heating public baths.

Based on the 70 to 90 million tons of slag they've found, environmental archaeologists have been able to estimate the ancient demand for wood. That amount of slag—the residue of the smelting process—represented 50 to 70 million acres of trees. Put another way: To smelt one ton of silver, Romans had to burn 10,000 tons of firewood.

As the Greeks did earlier, the Romans carried out military strategy based on timber: to get forests, they conquered them.

Deforestation came with an evil friend called erosion that did a real number on the physical topography of water-thirsty Mediterranean lands. Instead of a single environmental disaster, the Greeks and Romans faced a slow-motion cascade of catastrophe. There was growing damage to agriculture, since much good soil was lost to runoff. Springs, streams, and rivers were ruined by that uncontrolled runoff. Microclimates were altered as vegetation disappeared. Wetlands expanded, enabling the spread of malaria-bearing mosquitoes.

Attempts at reforestation were pitifully few. During Egypt's Greco-Roman era, the government deployed a nationwide tree-planting project. The seedlings, started in government-run nurseries, may have resembled the monoculture tree plantations of modern times. They supplied local wood by making good use of wastelands, open terrain on royal estates, and the banks of rivers and canals.

Equally farsighted were the measures taken here and there by individuals, rulers, and governments to protect woodlands—and, at rare times, the wildlife dependent on them. Cyprus, famed for its trees, did have certain dynasties of rulers who protected the forests for centuries, regulating timber cutting and engaging in selective harvesting. These forces for preservation and conservation, however, proved too minuscule to prevail.

Today's Mediterranean populations pay the price for all those centuries of shortsighted excess and waste. Cyprus, the island named after copper, mined its red-gold wealth for 3,600 years. Although reforestation has been carried out in modern times, more than 4 million tons of slag still sit on Cypriot lands, a somber reminder of millions of acres of long-lost woodlands.

RIVER OF HEAVEN

In the seventh century B.C., the Greeks looked into the star-heavy skies above them; stirred by the million points of light they saw, they wrote poetry about "that shining wheel, men call it milk." They were describing the Milky Way; and their word for "milkiness" became our word *galaxy*.

Galaxy then didn't mean what it means today: that is, an enormous cluster of stars, often organized in a spiral. Since early stargazers weren't able to perceive the immense depth of space and the relative distance of stars, they referred to the clusters of stars they saw as constellations, not galaxies.

The Greek myths of how the Milky Way got splashed across the heavens represented another chapter in the life of Olympia's most dysfunctional couple, that bellicose whiner Hera and her husband, Zeus, the wandering-eyed king of the gods. Not content with the progeny Hera kept producing, Zeus decided to father a powerhouse son with a mortal woman. As anyone who's read Greek myths knows, on the mean streets of Olympia it never worked out well for the gals who dallied with Zeus. Hera, a more ancient deity than Zeus, invented jealous-wife radar. She invariably found out and turned Mr. Thunderbolt's lovers into a goose or a tree.

Frankly, I'd rather raise peacocks than deal with a kid like Heracles.

This time around, Zeus had his eye on a fetching gal named Alcmene. To outwit Hera, the head philanderer used his superpowers to take on the form of Alcmene's husband and had a rip-roaring three-night romp with her. (The cuckolded husband got one of those convenient mistaken-identity "we need you out of town" errands.)

In due time, Alcmene gave birth to a child named Heracles (meaning "the glory of Hera," another puzzling non sequitur of Greek myth). To keep her as-yet mortal newborn from being zapped by Hera, the new mother pretended to give him up for adoption by exposing the babe in a stony field outside the city of Thebes. Meanwhile, Zeus persuaded his daughter Athena to take Hera, who was breast-feeding an unnamed offspring at the time, for a casual stroll.

When she spotted the kid, Athena did a fake double take. Picking up Heracles, she said, "Wow—what a wonderfully robust abandoned child! Must have been a crazy lady to let this one go." Significant pause; then Athena added, "Wait a minute. You have milk—why don't you let the poor little thing suckle?"

Hera, busy contemplating new ways to torture her husband, absentmindedly agreed, handing off infant number one and taking the foundling child in her arms. The baby boy being Heracles, he latched on with such powerful, painful suction that Hera threw the baby down. As she did, her mother's milk flew across the heavens and became the Milky Way.

Crying out, "You young monster!" Hera stomped back to Mount Olympus, looking forward to another vicious fight with Zeus. Meanwhile, Athena dusted off the baby and returned him to his mortal mother, announcing, "The kid's immortal now—but you'll still have to guard him and bring him up."

Heracles went on to strangle snakes, clean stables, and kill lots of people and animals in ingenious ways, from head-butting to assault with a lyre. He grew up to be a demigod, enthusiastically worshiped by both Greeks and Romans, who spelled his name Hercules.

Later, it was said, some of the drops of Hera's star-milk formed curds in the heavens; instead of cottage cheese, they in turn became planets.

The Greeks were insatiable when it came to myth spinning; Roman writers and poets also added to the tangle. In Latin, the Milky Way became Via Lactis but was also called the Heavenly Girdle by the Romans. There were almost as many variants of the Milky Way myth as there were stars.

Engaging myths notwithstanding, Democritus and a few other perceptive thinkers recognized the Milky Way as a vast assemblage of stars, very far away.

STAR STRUCK

Once upon a time, the skies over the lands and islands called Greece and Italy had a translucent quality. At night, the sky wrapped the earth in a blanket of dark lambswool, richly embroidered with pearls and bangles of silver. The people below those enchanting heavens spent a lot of time gazing at them. For those who depended on agriculture, it was second nature.

Today we briefly glance overhead to assess the weather or admire an especially rich sunset. Sky watchers of old, sitting on what they thought was the firm and stationary earth, felt an intimate connection with the objects moving with slow majesty before them. They called them fixed stars (*aster* in Greek, *stella* in Latin, from which comes *constellation*) or wanderers (*planete* in Greek, *stella errans* in Latin). They gave them names, powers, and stories. Great stories. For the watchers, the panorama of the night was a long-running celestial series about the quarrels of the gods, the feats of heroes, the quest of Jason, the flight of Pegasus.

Primetime viewing, complete with 3-D and a really big screen

And the Magnificent Seven, vagabonds all: sun, moon, Mars, Mercury, Jupiter, Venus, and Saturn. Each had its own color, number, aspect, and day of the week.

Venus got rapt attention for its brightness and its Jekyll-and-Hyde way of becoming both morning and evening star. If you were very patient and lucky, you might see the planet gleam bright enough to cast a shadow—or hang like a jewel against the daytime blue.

Pliny the Roman encyclopedist gave Venus a glowing write-up in his *Natural History*. Among other things, he said: "Below the sun revolves a very large star named Venus . . . Rising before dawn, it receives the name of Lucifer, as being another sun and bringing the dawn, whereas when it shines after sunset it is named Vesper, as prolonging the daylight, being a deputy to the moon. Its influence is the cause of the birth of all things upon earth; at both of its risings, it scatters a genital dew with which it not only fills the conceptive organs of the earth but also stimulates those of all animals."

Planet Venus as fertility goddess. That notion went back to Mesopotamian times, when the brilliant sky gem was called Inanna or Ishtar by other stargazers. It's the only planet ever thought of as a female deity. Like the moon, that other heavenly female, the sharp-eyed Greeks saw that Venus had a monthly cycle of waxing and waning.

In writing *The Iliad*, Homer had been the first to name the stars and constellations in print, but it wasn't until the fifth century B.C. that planets and star groupings were tied to specific myths. As they became more closely identified with deities or heroes, the heavenly bodies became physical representations of the divine.

Already worshiped throughout Greek-speaking lands as Aphrodite, Venus worship in all its aspects soon became tantamount to a franchise in Rome and other parts of Italy (notably Pompeii and Sicily). Rome's oldest temple to Venus was built in 295 B.C. A busy fellow named Quintus Gurges had vowed a sanctuary to the goddess once he finished fighting the Third Samnite War. A thrifty fellow, Quintus had promised merely to build a temple, not to fund it. Marvelously enough, a new source of funds bubbled up after the Roman treasury began to levy fines against women who'd committed adultery. The temple was built and dedicated in record time, indicating there were more than adequate numbers of guilty parties—and accusers. This temple's particular Venus was known as Obsequens, meaning "indulgent one."

Later, it got competition from the temple of Venus Erycina, named after a similar temple at Mount Eryx in Sicily. Sicilian prostitutes had taken to that Venus in a big way. That temple and the one in Rome became favored gathering spots for off-duty hookers. Annually, they held a wine-bibbing festival called Vinalia Priora, at which time call girls from the elite ranks to the humble brought offerings of new wine and other goodies.

Yet another Venus took up residence in Rome near the Cloaca Maxima—the Great Drain of Rome's amazing sewer system. Her temple, built in 39 B.C., was conveniently positioned to cater to the workers who cleaned the sewers and kept them in repair. Although the goddess originally shared billing with a water sprite, later she was known simply as Venus Cloacina, the purifier.

The list of Venus temples within Rome city limits went on for miles; so too did the number and variety of sites dedicated to Saturn, Mars, Mercury, Jupiter, Luna/Selene, and Sol/Helios, each serving a different segment of the population.

The Romans popularized the worship of these star-deities in all their aspects, but it was the imaginative Greeks who birthed the ideas. In addition to the heavenly bodies already mentioned, they saw and named forty-eight different constellations. If we think about constellations at all, we think of them as groupings of stars. Most of the stars in any given constellation, however, are separated by vast distances and often are traveling speedily away from one another. They just happen to lie in the same direction as we (and the Greeks of old) view them from earth.

To keep warm, it's likely that the sky watchers of long ago may have taken along some liquid refreshment. That might account for some of the wildly imaginative names the Greeks gave to the constellations. As they looked up and connected the bright dots, they saw action figures: scorpions biting bulls, Orion the hunter fighting unicorns and griffins, Heracles taking on bears big and little.

Because of our modern star illiteracy, Ursa Major and Ursa Minor are often thought of as synonymous with the Big and Little Dippers, but in point of fact they're part of the larger bear constellations. To make matters more complex, in the original Greek story, the bear was a dog. That helps explain why the "tail" of stars leading from Polaris, the North Star, is so very long. With the Greeks, of course, you always get more than one myth. The seven stars we think of as the Little Dipper they also called the Hesperides, the seven daughters of Atlas.

Our sophisticated world no longer thrills to Greek and Roman mythology, but we carry remnants of that starry-eyed romance in our language. *Disaster* still means "an ill-starred calamity." *Stellar* still signifies a star performer. And *asterisk* is our "little star," that twinkle of a symbol we use to alert readers to a footnote or another small surprise.

ANOTHER KIND OF LUNACY

Moonlighting used to be a lot more literal. It began when places such as Attica (the region that included Athens) and other parts of Greece and Italy were still rural, agricultural societies.

Early on, a notion sprang up that the moon created evening dew, vital to the nonirrigated crops in those days. Pretty soon farmers became convinced that the amount of dew—and crop growth—waxed and waned with the moon. From there it was inevitable that guidelines for planting, harvesting, and other rustic matters would take hold.

Agricultural authors such as Columella got into the act as well, writing books on planting for dummies. Nevertheless, most believed that the best advice was a father-to-son thing, carefully handed down through oral traditions.

Of mice and men and moonlighting

Farmers were warned to sow lentils before the twelfth day of the lunar cycle, as the moon increased, whereas beans needed to go into the ground the day of the full moon. Fruit-bearing trees were fussier; olives, pears, figs, and apples had to be planted on an afternoon during the dark of the moon, with absolutely no winds blowing from the south.

Moonlight rules for harvesting were complex, too. Grapes, if picked for raisin use, also needed a waning moon, whereas grapes for wine had to be picked at the waxing crescent for maximum juice. Both cultures ate huge amounts of onions, leeks, and garlic. For more refined palates who wanted less garlicky breath and milder flavor, those plant species were put in the ground and harvested in the dark of the moon.

A great deal of work must have been done by literal moonlight. Or no light at all. For instance, grapevines would be safe from mice only if they were pruned by the light of the full moon. Tree felling needed to take place between the twentieth and the thirtieth days of the lunar cycle. Manuring of crops and vines was done when the moon was dark, to kill the weed seeds.

Winemaking—and drinking—got its own lunatic set of laws. The must had to be boiled at night during the dark of the moon. Wine jars could be opened only during the full moon, so as not to sour or bring on a second fermentation.

These principles of sympathetic folk magic spread like dandelions to other farming operations. To ensure success, a farmer needed to dig ditches, castrate farm animals, salt down meat, set hens to laying, and shear sheep at the correct lunar phase.

Reaching their most florid stage, lunar precepts spread to nonfarm populations as well. For example, males with balding problems took care to have haircuts only when the moon was increasing in size, whereas hairier fellows could profit by the opposite tack.

In December and January, hardworking tillers of the soil relaxed a bit to pay respects to the earth's sole satellite (although they did not think of it in those terms). They held the Compitalia, a festival to placate the *lares*, household guardian spirits and the oldest deities of the farmlands.

In spite of the moon lore they meticulously observed, agriculturists didn't do much actual lunar worship. Their moon goddess, Luna, was a minor league deity, and so was Selene, her Greek counterpart. Instead, Roman farmers built shrines to Pomona, goddess of fruit trees. Or to that old farm favorite Sterculinus, god of manure spreading.

HOW TO BIRTH A BOUNCING BOY

You'd never catch an Athenian male saying, "Yes, it's true—we're pregnant." It'd be even less likely to hear him warble, "We don't care if it's a boy or girl, as long as it's healthy."

As in other cultures, boys were the longed-for outcome. Greek ancestor worship required a son, since daughters would marry into another family. (Unmarried daughters—a father's nightmare—didn't figure into the equation.) In addition, sons were counted on to support their parents in their old age. Back when the notion of pensions or assisted-living facilities would have elicited incredulous laughter, parents counted on sons to do the heavy lifting for them—should they be lucky enough to reach old age.

Couples dealt with the issue of undesirable offspring in several ways. The most heartbreaking is illustrated by this brief letter written by a husband abroad on business to his expectant wife: "If the child is a boy, rear it. If it's a girl, expose the babe." At Athens and throughout Greece, abandoned infants were laid outside temples, city walls, and other public places. Passersby could adopt them; if they chose, the "adopters" could raise the infants as slaves, later selling them or fielding them as prostitutes.

The Spartans were icily eugenic, too: each child born was examined by the council of elders for deformities, illness, and other problems. Only healthy newborns were allowed to survive. The Spartans, however, did not pursue a policy that favored male babies over female ones.

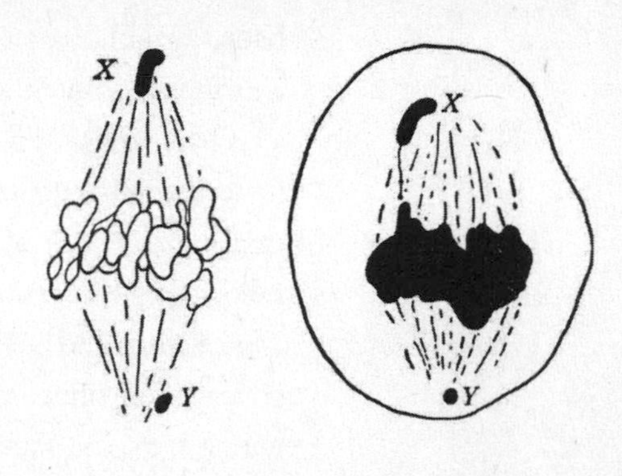

Two hundred weird ways to beget XY offspring

These harsh measures notwithstanding, most couples were deeply attached to kids. Rather than expose infants, they went to great lengths to use other methods, often resorting to magical (and decidedly unromantic) means to conceive offspring of the desired sex.

One finicky procedure had to be done prior to intercourse. Since males were thought to be generated from the right side, the putative dad manfully tied up his right testicle. Empedocles, a philosophical genius but out of his depth when it came to sex, had his own gender-begetting theories: males were created in the warmer part of the womb, he said. Due to their toastier prenatal environment, men turned out darker, stronger, and shaggier than women.

Other sex experts asserted that a baby's gender depended on whose semen predominated. Hippocrates, Democritus of Abdera, and Aristotle, among others, believed that there were male semen and female semen. When girls managed to get born, obviously the "stronger" male elements had been overpowered by an inexplicable abundance of inferior girlish sperm.

In addition, Aristotle was dead certain that girl babies tended to be born to younger women and to those nearing the end of their fertile lives—since in the first instance "the vital heat wasn't yet perfect," and in the second "it was failing." There being no gender head-counts in his day, his opinion went unquestioned. Aristotle supported another strategy, suggesting that male babies would predominate if couples would just remember to copulate during a nice strong north wind.

Other relied-on methods for boy babies didn't hinge on the weather, instead making use of plant materials. Right after sex, the female partner had to drink a mixture of raisin wine mixed with juice of a male parthenion plant, followed by eating the plant's leaves cooked in olive oil and salt. Certain plants were also said to lure the "baby on board" outcome away from girls: satyrion, crateagonon, phyllon, and orchis. Sometimes the roots or seeds were ingested; with others, the plant did its magic once crushed leaves or roots were applied to the genitalia.

Certain animal parts were also favored to conceive male children, rooster testicles or hare's womb being favorites. According to the Greek writer Plutarch, if at the time of conception a woman sat down to a nice meal of roasted veal with a side of aristolochia, she'd deliver a boy. It's more likely she would have expired: the aristolochia plant, popular long ago for a variety of maladies, was and is poisonous.

Sympathetic magic was at the heart of much advice, some so imaginative it's hard to fathom where it came from. A sampling: Men, look out if you're bedding your wife and she's got a black ribbon around her left foot—she'll conceive another useless female. Ditto if she starts knocking back cold water. White ribbon on her right foot? Full speed ahead for a boy.

The breasts of an expectant mother came in for close scrutiny during pregnancy. If they turned down, a girl was on the way. If both turned up perkily, get out the cigars. Should one mammary lose its fullness, it meant the loss of a boy or a girl. This odd belief had its roots in the nearly universal notion that the uterus itself had matching cavities, left and right.

Millennia before ultrasound took the guesswork out of boy-or-girl prognostication, parents did the best they could with what they had to work with, such as the advice given by Galen, one of Rome's high-powered doctors, to parents longing for boys. Instead of wearing perfume and a naughty nightie, a woman should smear her body with goose grease and resin from the terebinth tree for two days running, he said. A male child was sure to be the result—if, that is, she could persuade her male partner to get close enough to cohabit the next day. How Galen kept a straight face while saying this, we'll never know.

SECTION II

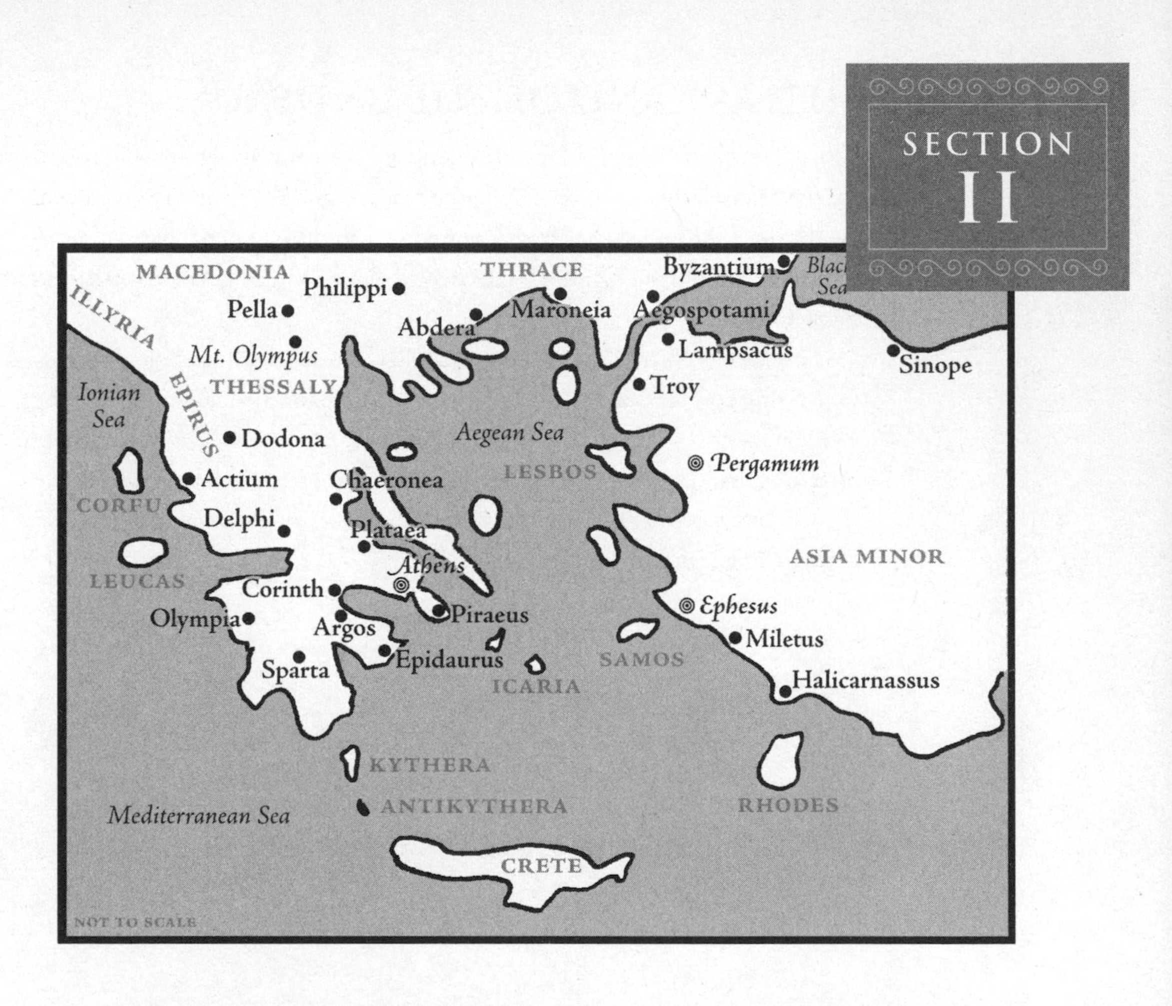

Greece & the Greek Islands

HELLAS, LAND OF THE UNDEAD

The nervously imaginative Greeks of long ago were not about to wait around for the Transylvanians to come up with vampires and werewolves. No indeed. By the time of Homer, they had dreamed up a nightmarish panoply of monstrous animal-human combos, netherworld deities, undead personalities, and bogeymen.

Greek pioneers in the supernatural could claim another first: bogeywomen. Garnering top fears among adults: the snaky-haired Medusa, one of the Gorgones trio. Her fangs and face were so hideous that the Greeks put them on their shields, hoping to turn enemy soldiers into stone—or at least unman them. Her image was also popular on amulets to ward off the evil eye, another high-polling fear among the Greeks.

The Gorgon sisters—we not only terrify, we do 20K events!

Everyone paid reverence to Hecate, the night-terrors queen of the phantom world. The Greeks often swore legal oaths in her name. Her domain became the crossroads, where once a month "Hecate's suppers" of dog's flesh were offered to placate her and as a rite of purification. Because of Hecate, Greeks young and old were afraid of the dark—and often missing small pets as well.

Hecate's henchwomen—Lamia, Mormo, and Empusa—did most of the everynight skullduggery in terms of haunting, flesh eating, blood sucking, and so forth.

Lamia claimed title to being the quintessential bogeywomen, a giant shark-like ogre who stole children and ate them. Greek parents often threatened naughty kids with a visit from Lamia; that, however, didn't stop several real-life women from proudly taking her name. One particular Lamia, a flute player whose real talents ran to playing men for pay, so entranced the Macedonian king Demetrius that he formed a liaison with her in the middle of a sea battle.

Mormo, doomed to play second fiddle to her sister specters, had been a queen in a prior life who'd lost her children. Due to that trauma, she had turned into a bloodsucking monster. In later Roman times, parents often dropped Mormo's name to terrify their kids.

But the seriously undead entity was Empusa, a malevolent vampire who sported one donkey leg and one brass leg. To have any sort of social life, this specter took over the bodies of living women from time to time.

One often-told anecdote about Empusa involved Menippus, a good-looking twenty-five-year-old who studied with famed philosopher Apollonius of Tyana. As Menippus walked along a lonely road outside Corinth one afternoon, a dainty woman with exotic looks deftly intercepted him, professed her love, and invited him to her home in the suburbs nearby. "I'll sing, there'll be wine—you'll be the only guy there," she said. Ditching his studies, Menippus kept the date; after a rapturous evening, the two became an item.

At length his philosophical guru, noticing the pale, baggy-eyed state of his pupil, warned him about the girl, throwing broad hints about the high failure rate among vampire marriages. Being in the throes of young lust, Menippus went ahead with plans to wed.

At the wedding breakfast, his worried teacher showed up, determined to carry out an Empusa intervention. Once Apollonius started challenging the reality of the golden goblets on the table and the servants attending the meal, both goblets and servants began to flutter away like the bats in a bad Bela Lugosi movie. After an interrogation by the senior philosopher, Empusa finally 'fessed up and admitted she had Menippus on a high-carb diet in order to devour his body—and not in an X-rated way. As she put it, she had a delicate constitution and needed to maintain a strict regimen of young hunks with pure blood.

Werewolves in ancient times got their share of glory, too—the best-known being a fellow named Daemonetos. As a Greek athlete from Arcadia, a wild and wooded region of Greece, he naturally took part in the local annual festival to Lycaean Jove—the highlight of which was the sacrificial barbecue of a young boy, followed by a no-host bar and hors d'oeuvres made of the minced entrails of the recently expired. After this particular carpaccio tasting, Daemonetos found himself getting extensively hairy and was obliged to take up werewolfing for a decade.

Disappointingly, the gone-with-the-wild lifestyle didn't offer enough drama, and in some mysterious fashion Daemonetos was restored to young manhood. Realizing there was no time to waste, he took up boxing and trained feverishly for the upcoming Olympic Games. He returned to Arcadia a winner, able to partake for free in all public feasts henceforth—but hold the *Homo sapiens* crudités, please.

Today's sports channels interviewers would sell their souls for a story like this, since not that many Olympic athletes have overcome the tragedy of a werewolf handicap.

THE FIRST SURROUND SOUND

If it weren't for Greek scientific sophistication about acoustics, we wouldn't have the Hollywood Bowl. Around 700 B.C., Terpander showed up, a Spartan who supposedly organized the first school of music, followed by

The theater at Ostia, one of numerous venerable venues with superb acoustics

that lyre-strumming philosopher Pythagoras, who created a branch of musical science called harmonics a century later. Inspired by their example, other brainy types developed the science of acoustics. Their greatest feat? Discovering ways to enable sound to travel from an unamplified actor to his audience. (Actresses didn't come on the scene until Roman times, so Greek actors played both genders.)

How did these unknown geniuses of theater acoustics figure out surround sound? To find answers, architects and archaeologists have studied ancient sites extensively—aided by the masterpiece theater at Epidaurus, a prime example of acoustical perfection, which has survived the millennia brilliantly.

At first, researchers assumed that the sculpted bowl shape of these outdoor facilities, plus the angle at which the tiers of seats were positioned, channeled the sound. Some also argued that wind direction played a role. At Epidaurus, for example, the prevailing winds tended to blow from the stage toward the audience—and still do.

But the drama-mad Greeks had other acoustic tricks up their sleeves. One principle developed by Greek architects and engineers and later written about by Pollio and Vitruvius was the notion of harmonics amplification. (Present-day sound engineers have tried without success to duplicate their methodology.) In small or medium-sized theaters built of stone or marble, a dozen niches were spaced equidistant from each other about halfway up the rows of seating. Within these niches were placed vases or vessels made of bronze. Not just any vessels, either; these were created to resonate at musical intervals, from fourth, fifth, and octave all the way to double octave. When actors emoted or the chorus sang, the sound traveled to the vessels, which resonated, supposedly enriching the words spoken or sung.

In larger theaters, architects incorporated three horizontal rows of niches, which used bronze vessels tuned to three different Greek musical modes. These acoustic enhancements may have enabled audiences of fifteen thousand or more to hear an actor's whisper in the back row. (A whisper measures 20 decibels, normal conversation 65.)

As if that weren't sophisticated enough to baffle the average Hollywood Bowl goer, another secret of exquisite acoustics remained unsolved until 2007. That was when French researchers reexamined the rows of stone seats at Epidaurus, which are smooth marble on top but corrugated on the vertical—a nuance long assumed to be part of the visual design. Instead, it was discovered that the corrugation acted as baffles for the sound, blocking low-frequency noises, such as crowd murmur, while transmitting higher-frequency sounds, specifically the voices of actors and chorus. Technically speaking, the frequencies up to 500 hertz were held back while those over 500 hertz were allowed to ring out.

As time went on, theaters and other structures designed for audiences were often built of wood. In these venues, architects didn't use bronze vessels, since wood was much more resonant. To amp up their sound, actors would turn toward the wooden folding doors at stage left or right when they began to sing in a higher key.

Although it's often asserted in print that the theater of Epidaurus was unique and that the Greeks could never replicate its acoustic perfection, that is far from certain. The works of ancient authors mention the names of various theaters that employed the bronze resonating vessels and were regarded as possessing superb acoustics. In addition, while it's probably true that the physical remains of Epidaurus are more intact than those of other theaters, it's equally true that Greek theaters from Aspendos in Turkey to Taormina on

Sicily to Ostia near Rome still boast marvelous acoustics, as modern audiences can testify.

Besides Pythagoras with his mathematical harmonics, other Greeks and Romans studied the physics of sound. Vitruvius, an architect-author who wanted to describe the acoustics of buildings, wrote, "Voice is a flowing breath of air, perceptible to the hearing by contact. It moves in an endless number of circular rounds, like the innumerably increasing waves which appear when a stone is thrown into smooth water, and which keep on spreading indefinitely from the center unless interrupted by narrow limits, or some obstruction." He actually lifted the stone-in-a-pond metaphor from a Greek philosopher named Chrysippos, long dead and thus in no position to sue; and other bits from Archytas, another early Greek, who produced a now-lost work called *On Audibles*.

The mathematician Euclid, whose famed work on geometry is still dreaded by generations of students, described sound more fully. Calling it a movement, he described the pitch of a tone as depending on the number of vibrations (or frequencies) in a given span of time. He put acoustics on a true scientific footing.

But the ancient world danced to other tunes besides theory and logic. Long-ago sounds could also have lucky or unlucky aspects. Whenever Greeks and Romans sacrificed animals at the outdoor altars near their temples, they always hired a musician, usually a player of the aulos, a clarinet-like instrument. He wasn't there, however, to contribute a reverent mood or to cheer up the proceedings. The musician wailed loudly away on his aulos to drown out other sounds. Any inappropriate noise, from a bystander's coughing spell to the bellow of an indignant ox getting its throat cut, was inauspicious and would require a do-over.

Speaking of inauspicious sounds: in recent decades, the gracious relics of ancient Greek and Roman theaters and amphitheaters have become venues for rock music and dance events. Since such concerts use amplifiers and loudspeakers, the heavy bass frequencies, coupled with ecstatic dancing, jumping, and other concertgoer kinetics, put an unprecedented strain on the stone and mortar of these monuments. Terrific ambiance for the fans, but for these irreplaceable buildings—bad vibes. Literally.

WIRED FOR HEAVENLY TUNES

Big-sky thinker Pythagoras firmly believed in a dynamic link between music, mathematics, and the heavenly bodies. Calling it "the music of the spheres," he saw it as the universal harmony produced by their tones. Since to him ten was the perfect number, there had to be an equivalent number to fit his philosophy. This required imaginative tweaking: Pythagoras had to dream up a counter-earth and two other heavenly bodies besides the seven he could see.

This crowd's getting antsy; time to play a slow-tempo Dorian number.

Greek music began with seven-stringed instruments like the lyre and the kithara, which a player would tune first with two strings, one octave apart, then build a sequence of notes using intervals such as fourths and fifths. The human ear can peg the intervals of these vibrating strings quite accurately. After discovering these major concords, Pythagoras developed proofs of their mathematical relationships. When played, the notes of an octave travel on a sound wave (or frequency) of 2 to 1, the sound waves of the interval of the fourth at 4 to 3, and the sound waves of the fifth at 3 to 2. All of these bits of math music were (and are) highly pleasing to the ear, whether played simultaneously or in succession.

The Greeks didn't use sharps or flats, instead incorporating quarter tones into their musical range. The tetrachord—a series of four tones with quarter-note intervals—was a key element in their system. They also devised two systems of musical notation, with one for vocals and another for instruments like the aulos, an early reed ancestor of the clarinet. These differences, along with the fact that only a few scraps of written music have survived, have made it difficult to reproduce what must have been a huge spectrum of Greek music. As with the other creative arts, Roman singers and musicians essentially studied the Greek system and adopted it. Little is left of their musical legacy, either.

From earliest Greek times, music provided a live soundtrack in peoples' lives. Storytellers sang their tales; poets warbled their words. Greek theater used choruses, dancers, and musicians; at festivals like the Thargelia, choirs of fifty men competed, along with elaborately dressed soloists on the kithara and aulos. Athletic festivals from the Olympics to the Pythian Games also held music competitions.

Men exercised to music, labored to it, went to war with it, dreamed of it.

In one of his dreams, Socrates was ordered to make music; once awake, he wrote a poem to accompany it. Women spun wool to music, got married and buried to its strains, sang it during festivals. Nearly everyone learned to sing or play an instrument as part of growing up.

Then as now, one generation excoriated another about musical tastes. Plato, for example, complained about "the pernicious tones corrupting our young people."

In the communal setting of his philosophical community in southern Italy, Pythagoras also taught that music could be used as a tool to heal, to soften anger, or to awaken patriotism. He began each day by playing lyre music for his followers. "Attuning souls," he called it. He also prescribed different modes of music for almost every ill.

He and the other Greeks organized music according to modes, which we would call scales or musical forms. Some—such as martial airs—correspond to modern genres or are still in use. Executed at a peppy pace, the Dorian mode got troops excited in a manly way before battle. It could also be adapted to a sober tempo to quiet crowds. Some Celtic musical pieces, such as "Greensleeves," and "Eleanor Rigby" use the Dorian mode.

In contrast, the Lydian mode was dreamy, even seductive at times. When played in high-pitched, plaintive ways, the Greeks used it for lamentation. These days it shows up in soundtracks and video games.

When played fast and wild, the Phrygian mode could induce religious frenzy; it invariably accompanied the orgiastic rites of the goddess Cybele. Slower and softer, the Phrygian was employed to heal such ailments as mental depression and skin disease. The Phrygian is still around; when performed in the dominant scale, it produces the characteristic chord changes of flamenco music.

The Greek art of the Muses had yet another complexity: tuning scales, called nomes. The most fantastic stories arose to explain their names. The Chariot nome, for instance. Eight reasons evolved for its name, some plausible (it imitated the high thin sound of a chariot's wheels), some weirdly original (the music played while a stallion covered a mare).

Since those long-ago centuries, researchers have made a number of discoveries that confirm what the ancients knew about the positive powers of music. Chronic pain and depression, for example, are now treated with applications of music. The field of kinesthesics has shown that music played at the speed and rhythm of the human heartbeat confers health benefits. The use of CPR in life-threatening situations has been shown to be more effective if done at 103 compressions per minute—which uncannily happens to be the exact tempo of that old Bee Gees tune "Staying Alive."

A neuroscience of music is emerging, one that explores how deeply music is wired into all of us. One of its findings? That the brain, while listening to music, releases dopamine, a neurotransmitter linked to pleasure. Another finding: that the cerebellum has a direct connection to our ears, producing some of our emotional responses to music.

Long-ago music lovers knew that as well. Writing to a relative, Pliny the Younger once said the following about his wife: "She even sets my poems to music and sings them, to the accompaniment of a lyre. No musician has taught her, but love itself, the best of instructors."

PHILOSOPHICAL FAVA PHOBIA

Even in his lifetime, he was called a miracle man and god-like. Awestruck groupies tried to sneak peeks at Pythagoras' supposedly golden thighs, a sure sign of divinity. After his death, rumors flew that his mother, Phythais, had hooked up with the god Apollo to produce her offspring. Alive or dead, the vibe around philosopher-teacher Pythagoras was celebrity-frenetic. (Like so much else, Greeks invented *frenzy* and its spinoff, *frenetic*; they come from *phrenitis*, "madness" or "delirium of the brain.")

Born around 582 B.C. on Samos, the son of a gem engraver, Pythagoras soon vagabonded off to Egypt, Persia, and other sites of esoteric learning, studying with sages, gathering secret lore, and getting initiated into mystery religions such as Greek Orphism.

Like the followers of Orphism, Pythagoras came to believe in reincarnation and the transmigration of souls. (Truth be told, at times he could be a real snooze, prattling on about his former lives as a fisherman and one of Jason's Argonauts.)

He moved to Crotona in southern Italy to found his philosophical school. His teachings and charismatic personality soon attracted hundreds of followers—and he made a special effort to encourage the participation of women, who were traditionally left out of such opportunities.

The founder ran his school on a two-tiered system. In the three-year "Pythagoras lite" program, participants only got to hear lectures. In contrast, the mathematics students endured a five-year probation and vow of silence, their property meanwhile managed by Pythagorean officials. At the end of probation, disgruntled students, if any, got their assets back plus interest, the world's first (maybe only) philosophical guarantee.

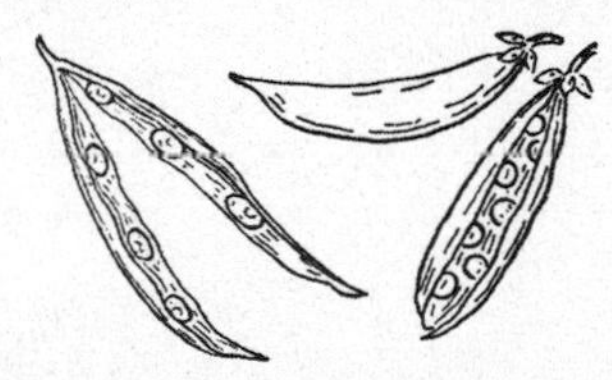

Decoding the Great Bean Taboo (and much else) of Pythagoras

The regimen was strenuous. Besides the study of numbers, geometry, philosophy, and holistic medicine, everyone took part in group exercise and ate a vegetarian diet. Pythagoras had his contradictions. Big on nutrition and an avowed vegetarian, he also coached an

Olympics winner by putting him on an all-meat diet. He also bragged about his knack of communicating with animals wild and tame, including an ox he once persuaded to swear off green beans for life. (A hint of what was to come? Perhaps.)

On many issues, he took unusual stands compared to other teachers. He was for ethics in business, against abortion and suicide. Conservative when it came to sexuality, Pythagoras insisted on minimal hanky-panky among his flock. Coitus was for procreative purposes only; his advice was to take up the lyre or take a cold shower. Despite his hard line on sex for other people, Pythagoras fell for a local teen named Theano and married her when he was fifty-six. In short order, they began producing a houseful of young Pythagoreans.

The man wrote copiously, although nothing remains that can be securely attributed to him except for a few sayings. As teacher and mentor, Pythagoras displayed a rational and compassionate mind, which makes it even harder to understand his Great Bean Taboo.

Long before Pythagoras, beans were viewed as magical. Taking beans to an auction was thought lucky; soothsayers put beans and salt in front of their customers before divining; legumes were offered to the dead at certain Roman festivals. But beans were also condemned as unlucky; finding a white bean or smelling the blossom of a broad bean was a sign of imminent death.

When Pythagoras announced to his flock his own ban on the fava bean, he gave at least five reasons for abstaining. "Beans produce flatulence and interfere with the clear functioning of body and mind," he said. After that, the logic started to unravel. "Beans make women sterile," he proclaimed. "Oh, and they look like genitals. Or the gates of hell."

Reason number five, however, was the sublime head-scratcher. At the beginning of the world, Pythagoras taught, men and women sprouted from the same putrefaction. "Eating a bean, then, is like eating the head of your parents," he asserted. "Crush a bean with your teeth and put it in the sun for a while. When you return, you'll find that it smells just like a murdered man."

Imaginative and puzzling claims indeed, from the same thinker who reputedly revealed the secrets of the hypotenuse.

What was at the heart of his beans-are-murder taboo? One clue has emerged from modern research. Scientists now know that the red blood cells of certain individuals lack an enzyme, G6PD, needed to break down a substance in fava beans called peptide glutathione. People who are G6PD-deficient may get dire reactions to bean pods and even to the plant's pollen, ranging from hemolytic anemia to jaundice, high fever, and death. Pythagoras, it's hypothesized, may have been a secret sufferer of favism.

There are still other chapters being written in the fava bean saga. Some years ago, researchers in Africa uncovered a relationship between sickle-cell anemia and malaria, showing how people who carried one gene for sickle cell (and were largely healthy, as opposed to people who had two such genes and developed severe disease) were more resistant to malaria, endemic in the same regions. Exciting research on G6PD deficiency now hints that this condition, like carrying a gene for sickle-cell anemia, may have survival value against malaria.

For millennia around the southern Mediterranean, consumers of fava beans, including those who have a negative response to them (an inherited trait genetically passed through females), have also lived in mosquito-infested, malaria-prone areas. The negative impact of having G6PD deficiency may be offset by the positive value of enhanced protection against the greater of two evils, malaria.

Oh, and the final irony of Pythagoras' bean ban? Centuries after Pythagoras' death, a number of his more fanatical followers honored the taboo assiduously, to the point of sacrificing their lives rather than tread on fava bean plants. Unlike Pythagoras, chances are good that none of them even had favism.

INVENTING THE ATOM, SPLITTING THE CREDIT

In the fifth century B.C., a duo of acute thinkers named Leucippus and Democritus came up with the notion that all matter was composed of tiny units that were indivisible. For that unit, they coined the word *atomos*, meaning "uncuttable."

It's hard to imagine the intuitive leaps and bounds it took to come up with this concept, given that these brainiacs had no electron microscopes, no federal grants, no way to test their hypotheses—only their hyperactive gray matter.

And it was bold, since the notion of atoms ad infinitum in the universe pretty much squashed the possibility of the Olympian gods taking any credit. As Democritus declared in print, "Nothing exists except atoms and empty space; everything else is opinion."

These days often labeled "the father of modern science," Democritus would have called himself a natural philosopher rather than a scientist or religious reformer. He came from Abdera, a Greek city-state in Thrace, many miles northwest of Athens. Folks from other parts of Greece used "Abderan" as a synonym for stupidity, which must have made locals chuckle, since a number of famed thinkers were born there.

His mentor and elder by thirty years or so, Leucippus was thought to hail from Abdera as well. A philosophical disciple of Zeno, he set up the atom as a first principle. Not all his ideas were on target; he also thought that the earth was shaped like a drum and that the stars were set on fire by the speed of their motion. Little more is known of him, except that his work provided the spark to fire up the beautiful mind of his protégé Democritus.

The twin contributions of these men were huge, yet their proximity to the truth about matter, the atom, and the empty spaces between the building blocks of the universe remained largely unexamined until modern times.

As a young man, Democritus traveled widely, studying with the Egyptians, the Chaldeans of Mesopotamia, the sages of India, and the Pythagoreans. Being the third son of a wealthy man, he asked for his inheritance in cash, which he spent on travel to these far-flung places.

To further educate his mind, he studied physics, mathematics, ethics, and the arts. At times, Democritus was troubled by visual hallucinations or too-vivid nightmares. To gain insights about his inner life, he conducted sensory deprivation experiments on himself. How? By spending an appreciable amount of time in darkened tombs. Maybe he craved the quiet (perfect for high-quality thinking) that he found there.

Once he felt ready, Democritus began writing. Lifelong curiosity, intellectual vigor, and an eighty-year life span allowed him to create a huge body of work. His grand themes: exploring what the world was made of and how the natural laws of the universe worked. His subject matter included celestial phenomena, the senses, colors, geography, rays of light, rhythms and harmony, medical regimens, and prognostication, along with lighter-hearted subjects such as painting and fighting in armor. He modestly called his master work *The Lesser Diascosmos*. Its companion, *The Great Diascosmos*, meaning "the great world system," was attributed to his teacher Leucippus.

Democritus, who was of Socrates' generation, also went to Athens to study with that great teacher, choosing to remain in the background. The ever-so-touchy Plato, whose beliefs and writings represented an elite worldview bolstered by unprovable notions about prime movers and the purpose of the universe, was grossly offended by Democritus' ideas—and, frankly, burned up at his popularity. Speaking of combustion: Plato actually tried to collect and burn all of Democritus' writings. This thuggish act was discouraged by several Pythagorean philosophers, who told him, "You're too late, dude . . . his books are all over the place."

Plato, Socrates' self-appointed biographer, maintained his grudge, never mentioning Democritus' work. The whole thing became a testy, long-running philosophical-scientific quarrel in Athens.

The intellectual hullabaloo later extended into the Aristotleian camp. As Plato's own prize pupil, Aristotle naturally defended his teacher—at first, anyway. Later he went his own way, blowing off many of his mentor's theories and (gasp) founding a rival academy called the Lyceum. Far worse, he had the gall to applaud the work of Democritus. Well, some of it, anyway. Being more empirical scientist than head-in-the-clouds moral philosopher, Aristotle applauded Democritus' insightful research into flora and fauna.

On the other hand, Aristotle hated Democritus' idea of atomic particles, lobbying instead for his beloved scheme of four essences (fire, air, earth, water), plus a fifth, an ether called quintessence. Strange as it may seem to us, Aristotle's point of view came to dominate scientific thinking for centuries. One compelling reason: a vast portion of his books and notebooks survived through time, whereas only meager scraps and quotes by Democritus did.

Plato may have panned the Abderan's book, but loyal locals loved it. It was read aloud in Democritus' hometown, after which the townspeople awarded him 500 talents (a juicy sum for his old age), plus bronze statues in his honor and the promise of a public funeral.

Despite his distance from us in time and place, Democritus comes across as an empathetic human being, not an ivory-tower elitist. When he got quite old and feeble, his sister worked herself up into a snit, thinking that her brother might die during the upcoming women's festival of Thesmophoria. Clearly that would ruin her plans; women of her day didn't get to party that often with other women.

Democritus reassured her, saying, "Go enjoy yourself, don't worry, I won't quit the earth just yet." During the three-day festival, he had hot bread brought to him daily, which he applied to his nostrils. Shades of Betty Crocker, it worked. A reviving quality in bread atoms, perhaps? Promptly after the Thesmophoria, Democritus departed, having had his fill of yeasty ideas and fresh loaves.

He left behind these words, among others, which may hint at how he felt at life's end: "The end of action is tranquility, which is not identical with pleasure, as some have understood, but a state in which the soul continues calm and strong, undisturbed by any fear or superstition or any other emotion. This I call well-being."

NECROMANCING THE STONE

A question that has haunted mankind for millennia: what does a restless ghost need to achieve peace? The Greeks thought they had the answer: the art of necromancy. By it, they meant learning secrets from the dead, although nowadays the term is more loosely tossed around to mean black magic involving ghosts or demons.

To communicate with their dearly departed, Greeks sometimes employed skull necromancy, as the biography of Cleomenes will illustrate. A boy in Sparta around 540 B.C. who would grow up become a great king, Cleomenes had a best pal named Archonides. For years, they hung out together, swearing those blood-brother vows about being friends forever and never keeping secrets from each other.

A nanosecond after Cleomenes became King Cleomenes I, he cut off his bosom buddy's head, preserved the thing in honey, and kept it in a large jar in his quarters. Whenever a battle loomed, he would have a tête-à-tête with Archonides. If he wanted to double-cross an opponent, he talked it through first with his erstwhile friend. When he was about to found the Peloponnesian alliance that would ultimately destroy the Athenian empire, he discussed his plans with his pal. It may come as no surprise to learn that the honeyed skull of Archonides agreed with everything he said.

Not all skull necromancers found it necessary to slaughter the owner of the skull in order to communicate. A few incantations, some magic rites, and the skull's owner could usually be counted on to visit the necromancer in his sleep and provide answers. For instance, after Emperor Nero had murdered his mom, Agrippina, then gotten rid of the evidence with an appalling lack of religious ceremony, her guilt-ridden son couldn't get any sleep. To quell the ghostly harassment from the mother he'd killed, he hired a Persian ghost specialist to summon her shade and appease her.

I'm Mercury and I'll be your soul conductor this evening. Newly dead spirits, follow me . . .

Beyond Necromancy 101 were supernatural feats that required a lot more legwork. The recipe for reanimation, for example. Long to bring a favorite corpse back to life for a chat? First assemble your ingredients, including fresh blood, the foam of a rabid dog, and the hump of a hyena. Pump them into the corpse while gently reinserting the soul—a tricky maneuver.

Both the Greeks and the Romans regularly scheduled appeasement festivals to keep ghosts at bay. The Romans liked things compartmentalized and tidy, so they held separate events for different categories of the dead. In February came the Parentalia, a spirit-world fest held at the tombs for the "good" deceased, generally family members who hadn't committed suicide or murder, hadn't died young, and hadn't been killed by lightning.

In May they held the Lemuria, more of an exorcism than a joyous time to remember dear old Granny. During Lemuria, each household appeased the really hostile and spiteful ghosts called lemures and larvae. At midnight on the ninth, eleventh, and thirteenth of May, the male head of the house conducted a ghost-busting ritual. Barefoot, making "Spirit be gone!" gestures, he moved through the house, spitting out a series of nine black beans as he went, without looking behind him. The duties required good hand-eye coordination: simultaneous spitting and gesturing and calling out, "With these I redeem me and mine!" Once the master of the house had made the circuit, he did a quick hand washing, some loud banging on brass pots, and a final "Spirits of my ancestors, get lost!" and he was through for that night.

There were three more occasions when spirits of the dead wandered footloose in Rome and other towns. On single days in August, October, and November, the mundus, regarded as an opening straight into the underworld, was opened. Normally the mundus (more like a large stone acting as a manhole cover) remained over a trench in the ground. While the ghosts were free to boogie about town, the activities of the living came to a halt.

The Greeks celebrated parallel rites, the main one called Anthesteria, when ghosts from the underworld entered the city of Athens. During the festival, the ghosts got a meal of mixed grains and then had to be chased out. To protect themselves against excessive haunting, Athenians put pitch over their doors and chewed hawthorn leaves.

Through ritual, ancient folks took care not to encounter ghosts, but they loved vicarious experiences of the supernatural. Their literature is filled with ghost stories, hauntings, revenge from beyond the grave, and grateful-dead tales.

One of the most touching is the much-repeated account of the shade that visited Brutus, the young aristocrat who was both a protégé of Julius Caesar and one of his assassins. Late one night, Brutus was sitting in a battlefield tent at Phillipi in north-central Macedonia, mulling his uncertain future.

An apparition appeared at his side. Brutus asked, "Who are you—of men or gods—and upon what business do you come to me?"

The eerie figure answered, "I am your evil spirit, Brutus, and you shall see me at the battle of Philippi."

Brutus kept his cool. "Then I shall see you."

And at Philippi, where the two largest Roman armies ever assembled did battle, Brutus did indeed meet his shade and with near eagerness go to join him.

RAIL SERVICE, MINUS THE TRAINS

Isthmus: a tongue-twister of a word, coined by the Greeks. All it means is a skinny strip of land connecting two larger pieces of real estate.

In ancient times, the most famous isthmus was the Isthmus of Corinth, an umbilical cord 20 miles long and 4 to 8 miles wide that linked the southern chunk of mainland Greece with its bigger northern half.

On it sat the rich city of Corinth, its location favored by the gods and by commerce, rich and gaudy and full of illicit pleasures. The sexual saleswomen of Corinth were known as "the colts of Aphrodite" for their frisky repertoires. It was the Las Vegas of Greece—without neon or slots, but Sin City all the same.

From time to time, Corinth was governed by tyrants, few of whom would have won nice-guy awards. Periander was a spectacularly cruel specimen who went down in history as a necrophiliac (yet another useful Greek term, used to describe those individuals with an erotic attraction to corpses).

Periander also had a gift for imaginative engineering. His city was already a busy shipping hub. Still, a shortcut from the eastern waters of Greece to the west, one that avoided the need to sail the dangerous waters around the big fat Peloponnesus, would fatten Corinth's coffers even more.

Staring at his isthmus, Periander saw that a canal would be ideal but knew that digging it would be a killer job. "What other options do I have?" he mused. And the muse answered: "A diolkos, you dodo."

Accordingly, in 600 B.C., he began to construct a peculiar trackway that began at the waters of the bay of Corinth and ended at the waters of the Saronic Gulf. Paved with hard limestone, the 4-mile-long, 20-foot-wide *diolkos* followed the gradient of the land. Its main feature? Two deep parallel grooves, about four and a half or five feet apart, running all the way through it. Instead of a roadway, Periander had created (for lack of a better English word) a railway. As the muse had hinted, in order to have a railway, the one thing you don't need is a train.

On the *diolkos*, a variety of goods, ships, and wheeled vehicles could travel. Boats and other vessels up to a certain size (a trireme 121 feet long was the maximum) were pulled from the water, then transferred to a large flatbed trolley whose wheels meshed with the grooves in the *diolkos*. To protect the ship's keel, ropes and other materials were lashed around the vessel, the whole affair pulled by oxen and/or human labor.

The biggest drawback? The limitations imposed by the size and power of ancient cranes. Although the bigger cranes such as the multiple-winch *polypastos* had masts and four technicians to work it, its lifting capacity was only about 6,600 pounds. Archaeologists working on Corinth-area investigations, however, have unearthed a mechanical device that they speculate may have assisted in lifting weightier cargo onto the flatbed trolleys.

Heavy items such as marble and timber would be offloaded from the ships they came on, placed in the *diolkos* cars, and hauled to the opposite end. Modern scientists have penciled out what feasibly could have been transported and what it must have taken to do so. As the *diolkos* was at the narrowest part of the isthmus, it took about three hours to get from one end of the trackway to the other. Larger objects such as a trireme required teams of one hundred to two hundred men to pull them—more men for the initial hernia-inducing tug, less once inertia was overcome. Smaller loads required Zeus only knows how many oxen.

Before long, the arms of Periander the tyrant got tired from all the patting himself on the back he did. His *diolkos* trackway was an immediate commercial success—he charged tolls, of course.

Over its nine-hundred-year life span, the *diolkos* would prove invaluable in times of war. This being ancient Greece, that meant nearly all the time. Periander wasn't around to see it, but the Spartans became repeat customers in their various conflicts with Athens. Later, big names, from Philip II of Macedon onward, used the *diolkos* as well.

In 31 B.C., right after Octavian, Rome's future first emperor, beat Marc Antony and Queen Cleopatra VII at the Battle of Actium, he used the *diolkos* to whisk half of his 260 Liburnian ships, his speediest vessels, across the isthmus. In that fashion, he was able to chase Antony back to Egypt.

Although the *diolkos* of Corinth would remain the gold standard for heavy loads, throughout Greece (and other rural regions around the Mediterranean) there were more modest transport projects called rut roads. Being a land of mountains and rough, rocky terrain, Greece never had a system of paved roads, as Rome did from imperial times. Instead, in hilly areas and along the heavily trafficked gateways to oracles and bigger towns, builders carved parallel wheel ruts into both paved and unpaved stretches. This allowed wheeled vehicles, from carts to larger wagons, to struggle over otherwise impassable stretches. Like modern train tracks, at intervals the rut roads had "sidings" that branched from the main road, allowing vehicles to pass.

The remnants of many rut roads are still around, puzzling visitors to Greek and Roman sites, who imagine that overzealous use by speeding chariots must have caused those deep impressions in ancient roadways.

DARING DRILLERS

Although such a notion wouldn't have played well in the dining rooms of loquacious orator types or politicos on the make, some Greeks chose to be boring. But it was what they bored, rather than who, that mattered.

For instance, about 520 B.C., the chief city on the island of Samos found itself running short of fresh water. Polycrates, the tyrant in residence, was pulling his hair out. (An aristocrat who seized power unconstitutionally, he was nevertheless a caring tyrant.) Someone suggested going after water from a fresh spring on the other side of the island.

But it took an out-of-towner from Megara, an engineer named Eupalinus, to pipe up with an audacious plan. "All that's needed is to dig a tunnel half a mile long or so, run some pipelines through it, and shazam. Fresh water, no worries."

Thanking their patron goddess, Hera, for her celestial help, the worried Samians gave him the bid, muttering to one another, "Do you think he realizes there's a mountain between the spring and the city reservoir?" Polycrates threw some money at the engineer, gleefully thinking, "If he pulls this off, will I ever look good. Then he can get rolling on the rest of my wish list."

Eupalinus' solution, while brilliant, was an immense trigonometry headache and engineering nightmare rolled into one. Nevertheless, he tackled the project with verve. Even though the surveying instrument called the diopter hadn't been invented yet (and wouldn't until Heron of Alexandria did so six hundred years later), this seasoned scientist knew how to carry out the job. It was ticklish, his plan of action. He set two crews to work on either side of Mount Kastro, drilling toward each other. The mountain was another unhappy obstacle; solid limestone, and the engineer had to bore through the base of a hill nine hundred feet high.

He didn't plan some grungy little drainage hole, either. What Eupalinus proposed were two tunnels, the upper one deliberately horizontal and high enough for men to move around in it for repairs and cleaning. The final product ended up about 6 feet high by 6 feet wide, tapering at the top to a V-shaped ceiling—all of it cleanly cut from solid limestone. Beneath it was a separate gravity-flow tunnel with a 2-foot-wide water channel, achieved by sinking a series of vertical shafts at regular intervals, then calculating the gradient from one shaft to the next. Thanks to the boss man's skill, the two crews met almost exactly on target.

The double-tunneled approach wasn't an original idea. Earlier ones had been carried out in Mesopotamia, Jerusalem, and Thessaly, and after Eupalinus, similar water tunnels would be installed in Syracuse, Athens, and Acragas.

But the order of difficulty and the magnificent execution of Eupalinus drew all the applause. Herodotus, among others, marveled over this tunnel, already eighty-plus years old when he inspected it. He declared it "one of the three greatest engineering works of all the Hellenes." It never made the Seven Wonders of the World circuit, but should have. Much of the marvelous tunnel can still be accessed today.

Twenty-five hundred years ago, there were no jackhammers, no earthmovers, merely iron pickaxes, stone hammers, and human muscle. How did they break rock in those days? Where there was room, a battering ram was used to break the matrix. To make tunnels in mine shafts, Greeks also used a method whereby they heated the rock with fire, then doused it with vinegar. Once the fire produced surface cracks in the rock, pouring vinegar (or any acid) into the cracks would make it split. Some modern experiments have found that water seems to do the job as well—but vinegar had mystique and the weight of tradition.

Brute force B.C.: *to make tunnels, men used small battering rams to break the rock matrix.*

Another well-chiseled underground tunnel had a far different feel. Located in the region of Campania, Italy, the land of hot sulfur springs and volcanic activity, the site of its opening was known through legend as a gateway to the underworld. Some years ago, two amateur archaeologists started poking around this volcanically active area of Avernus and Baiae. In the process, they hit pay dirt, discovering a labyrinth of spooky underground passages culminating in what they surmised was some sort of religious sanctuary.

They and other Indiana Jones hopefuls have had tremendous fun debating its possible use as a long-ago tourist trap, the headquarters for an Oracle of the Dead cult, a parking garage for aliens, and so forth. More mundane hypotheses have called it a service tunnel for a Roman bathhouse.

Whatever purpose it served, the maze is amazing. Running for more than 200 yards to a depth of 140 feet, it ends at a stream fed by volcanic springs. The solid rock tunnels are a claustrophobic 21 inches wide, their walls filled with hundreds of niches for oil lamps. Who was its Eupalinus? Was he Etruscan, Italian, or extraterrestrial? To date, no one knows, but anyone can play.

RECYCLING COLOSSAL WAR TOYS

The island of Rhodes produced a special breed of men, from the best rowers and archers to the longest line of Olympic victors. Another thing; they believed in cooperation, a rare trait among the quarrelsome Greeks. Their lush home, bigger than most Greek islands, grew wealthy from trading and shipping spices, amber, and wine via their commercial fleet. The Rhodians, being natural diplomats, tried to stay neutral to keep trade flowing, but it wasn't always possible.

By 408 B.C., locals could sense the ominous way the winds of war were building. Spearheaded by Olympics champ Dorieus, part of a local athletic dynasty, the islanders agreed to pool resources and consolidate their three main cities into one highly defensible entity. With masterly planning, on the north shore they built a city of wide, gracious boulevards, with an excellent water supply and a central sewer system centuries ahead of Rome. They surrounded it with stupendous stone walls, studded with higher towers at intervals. Even their harbor had ramparts and towers. That fortification would serve them well.

A century later, in the wake of the death of Alexander the Great, the world's toughest fighter was headed their way, and he wasn't carrying an olive branch, either. Demetrius I, king of Macedon but hoping to recapture Alex's empire, couldn't get enough of war or territory; neither could his dad, Antigonus, an Alexander-era general still swashbuckling in his eighties.

When enemy forces surrendered, Rhodians recycled the losers' siege tower into a colossal bronze sun god statue. Result? Tourism magnet.

Demetrius attacked the Rhodians because, he said, they'd allied with Ptolemy I, another of Alex's generals. The real reason: the distinct possibility of fabulous spoils. He had a loutish, piratical side; besides his own fleet of two hundred ships, more than a thousand other vessels, many of them belonging to full-time outlaws, followed his lead, hoping to feast on the leavings.

To isolate the island from its trading partners, Demetrius built a second harbor, which proved unworkable. Then his army of forty thousand swarmed over the island, building a camp near the city and soon managing to breach the walls. Another plan gone bust, since seven thousand supermotivated Rhodians slaughtered the intruders and made quick repairs.

Now Mr. Macedonia got serious, erecting a ring of siege towers around the city, including the largest siege engine the ancient world had ever seen. This wheeled monster, square at the bottom, narrowing at the ninth story, stood 140 feet high. Called *helepolis*, or "city taker," it was fireproofed with iron armor and mounted on iron-plated giant casters that moved in any direction. The thing may have weighed 125 tons; it took 3,400 men in relays to move this vertical forerunner of the tank.

Through portholes that opened and closed mechanically, soldiers fired flaming arrows, catapulted firepots and huge stones, and worked grappling irons and battering rams. One such ram was 200 feet long. (And yes, many of them did have ram's heads of decorative bronze on the business end.)

The Rhodians, meanwhile, weren't idle. First they freed and armed their slaves, giving them another sixteen thousand men. They set mines, fired on Demetrius' forces nonstop, and devised devious countermoves—such as flooding the area where a siege engine was getting in position, causing it to tilt and fall.

As the months wore on, Demetrius grew disgusted with the continued failure of his colossal siege engine—to say nothing of the bad press he was getting around the Mediterranean for his buccaneering approach. He was, frankly, shocked at the resources the Rhodians had been able to deploy. After one night's bombardment, he'd had his men collect the evidence off the battlefield. That night alone, the Rhodians had fired more than fifteen hundred catapult bolts and eight hundred projectiles.

At the end of year one, despite his three-to-one manpower advantage and light casualties, Demetrius decided to call it quits and inked a peace agreement with the islanders. Like other hubristic commanders in chief, upon leaving he proclaimed, "Mission accomplished!" He even left behind his siege engines, including the nine-story monster. Holding their breath at this stroke of luck, Rhodians agreed to remain neutral in Demetrius' upcoming war with Egypt.

After the Rhodians honored the fifty-four hundred who had died on their side, they took stock. Being good businessmen as well as die-hard fighters, they eventually held a huge garage sale of lightly used weapons of mass destruction. They also melted down all the metal plating from the city-taker siege engine, along with other metallic bric-a-brac.

With the proceeds from this recycling, in 302 B.C. they commissioned the Colossus of Rhodes, an offering of thanks to their island's patron, the sun god Helios. The striking 120-foot bronze statue faced toward the sea, welcoming arrivals at the harbor. Taller than the Statue of Liberty, it stood in a similar pose—not astride the harbor, as imaginative artists have portrayed it. Built in just twelve years by the Greek architect Chares, its high-tech assembly of iron struts and 13 tons of bronze was as ingenious as its financing.

The Colossus of Rhodes soon became one of the Seven Wonders of the ancient world. A mere fifty-six years after the ribbon cutting, however, the Colossus disappointed its fan base by snapping off at the knees and collapsing in a severe earthquake. Anxious to rebuild, city officials consulted the oracle at Delphi—but were warned not to go forward. As the Pythia (the female mouthpiece for the oracle) said, the sun god wouldn't like it.

Even in pieces, the fallen Colossus remained a Rhodes tourist attraction and profit center for centuries. Few men could get their arms around one of its thumbs; its fingers were bigger than most statues. Visitors posed inside or on top of its Goliath-sized body parts—then had quick-sketch artists capture their pre-Kodak moments.

And what of Demetrius, at one time the world's most feared general? After his failure to defeat Rhodes and the abandonment of his dud of a siege machine, people began to call Demetrius "Poliorcetes," or "city besieger"—a double-edged nickname used at every opportunity. Besieger, not victor. Who says that the ancient Greeks had no sense of humor?

AIRBORNE HITS AND MISSES

The myth of the rebellious teen Icarus and his dad, Daedalus, was among other things a classic Greek "mind your parents" morality tale. As the inventive father and son made their escape from Crete, Icarus wanted to show off his chops, flew too close to the sun, experienced a waxed-wing meltdown, and plummeted into the Ionian Sea.

Hardly anyone remembers what happened to his father, however. The answer? Daedalus flew on! After naming an island Icaria after his boy, the grieving craftsman continued his island-to-island flight path, ending up on Sicily. He dedicated his wings to Apollo, then became a big hit on the ground as a fix-it fellow for the local king. (The name Daedalus means "cunning worker"; he became the prototypical ideal of craftsmen and carpenters.)

The Greeks had a powerful interest in flight, as evinced by multiple retellings of stories such as those of Icarus and Daedalus, Pegasus the flying horse, the sun god driving his chariot across the sky, and other airborne myths.

But flight fabrication wasn't all mythology. The Greeks produced a cadre of brilliant cogitators who came up with the basic underpinnings of modern science, including baby steps toward achieving flight.

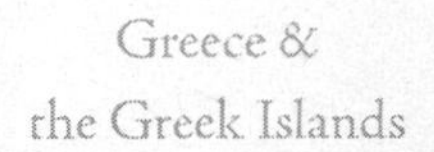

Like other aerial pioneers, the father-son team of Daedalus and Icarus encountered fatal design flaws.

That Icarian goal was one of many fascinating experiments carried out by Archytas, a now-shadowy figure seldom given much coverage in today's history books. A lifelong native of Tarentum, one of myriad colonies founded by the Greeks on the southern toe of Italy, Archytas lived about a century after Pythagoras and became one of that school's brightest teachers and mathematicians. Politically and militarily astute as well, this all-rounder was selected by his fellow citizens to be its *strategos* or ruling general seven times. When philosopher Plato sailed from Athens to Tarentum to drink more deeply from the Pythagorean school of thought, he went to the best—Archytas.

The Tarentum native ended up rescuing Plato from a nearly fatal pickle he got himself into with a Sicilian tyrant who wanted private lessons. Like Yoko Ono with John Lennon, Archytas' name thereafter became linked in second-banana fashion to the more celebrated Plato. Besides bailing out hapless celebrities held hostage, solving complex geometry problems, rendering mathematical proofs of the musical modes of Pythagoras, and writing treatises on acoustics and other topics, Archytas fiddled around with flight.

Around 400 B.C., he launched what's been called the first self-propelled flight device. He called it "the dove." As the Roman writer Aulus Gellius described it centuries later: "Archytas made a wooden model of a dove with such mechanical ingenuity and art that it flew; so nicely balanced was it, you see, with weights and moved by a current of air enclosed and hidden within it."

Some researchers think that the "hidden current of air" could have been a type of propulsion using jets of steam that another genius, Heron by name, later refined. Heron's array of machines included a variety of mechanical novelties propelled by hot air.

Archaeological finds from Greco-Roman Egypt of 200 B.C. provide another small but intriguing clue. Found in a tomb near Saqqara was a 7-inch-long wooden model, bird-like in form. Careful scrutiny by aeronautical experts has shown that its straight wings, tapered body and vertical tail fin more closely resemble an airplane or a glider. It glided easily through the air, causing some to theorize that it was a scale model of a flying device. Will we ever know if it—or others—got off the ground? In archaeology, discoveries that overturn current beliefs happen every day; stay tuned.

In another, more vicarious fashion, the Greeks and Romans did get airborne. Well before the classical age, Greeks were ardent pigeon fanciers, raising them as pets and for their abilities as racers and mail carriers. These strong flyers could achieve speeds of up to 90 miles an hour. A species of Italian swallow was also popular with the airmail crowd. What sort of messages got sent via pigeon post? Breaking news, such as the latest Olympic Games winners; lightweight stuff, including love letters; and highly strategic intelligence, useful to military operations.

For example, during the years-long manhunt for Caesar's assassin Brutus, Roman general Marc Antony found him hiding out in northern Italy and promptly besieged Modena, the city that sheltered him. Brutus, however, managed to communicate with his allies via carrier pigeon—and flew the coop to fight another day.

The oracle of Delphi also kept dovecotes and aviaries, ostensibly because such birds were sacred to that shrine's deity. Most oracular pronouncements were thought to be genuine, but when big political or military decisions hung in the balance, there may have been some creative fudging.

Take, for instance, the run-up to the first Greek and Persian war. Much earlier, a farsighted Athenian general named Themistocles realized that the only way his city-state could hold its own was with a fleet of warships. He persuaded the Athenians to build their first fleet, using the windfall riches that were emerging from their new silver mine at Laurium.

Before long, the Persians began to assemble a vast army and move against the Greeks. A worried delegation of Athenians went to the oracle at Delphi to seek advice from the Pythia. Her words of warning were dire; not only that, but when they got back to Athens, they found that a few statues of deities were seen to sweat blood. Now terrified, the delegation returned to Delphi, pleading for a prediction that offered a little hope. Strangely enough, her poetic phrases about "walls of wood," "divine Salamis," and "do not wait peacefully for the army gigantic" seemed to reinforce the advice of Themistocles, who'd urged his people to fight at sea, even though their fleet was new and untried. They did, and that unlikely battle was won, as the Greeks bottled up the huge Persian fleet and stomped them in the narrow straits between Salamis Island and the Greek mainland.

Perhaps Themistocles and the Pythia had great mental telepathy—or maybe a little bird told her how high the stakes were for the Greeks.

USER-FRIENDLY, WAY BACK WHEN

Until recently, only meager concrete evidence had surfaced in our time about the mechanical devices that ingenious folks of two thousand years ago might have used to calculate the movement of the planets and predict eclipses. In fact, modern historians and scientists have tended to shrug at the mentions of marvelous machines in ancient literature, since most of those describing such devices were neither scientists nor even living in the same century as the objects they were writing about.

That skepticism would change in the early 1900s, when sponge divers off Antikythera, a rocky islet near Crete, found the wreckage of an ancient ship, its 150-foot length still loaded with merchandise. As the first major underwater archaeology discovery, the find created excitement worldwide. Amid the life-sized bronze statues and glitzier artifacts, divers retrieved a mysterious lump of corroded bronze. Puzzled authorities thought it might be a mechanical artifact; an astrolabe, perhaps? Unable to view its innards before modern imaging technology, its function remained enigmatic for over a century.

Most scientists now agree that the Antikythera Mechanism, as it's called, represents the world's first analog computer. (Some argue it's more accurately called a calculator, since it does not appear to be programmable.)

Semantics aside, the device, dated between 140 and 100 B.C., is wonderfully sophisticated. It has layers of intricately meshed gears of different sizes. They represent the first known prototype of a differential gear, the principle that allows the wheels on your car to turn at different rates on the curves. Until this discovery, scientists thought that gear trains able to carry motion from one driveshaft to another were invented in the seventeenth century, not eighteen hundred years prior.

What purpose do these thirty gears serve? Together they mimic the movements of the planets, the sun, and the phases of the moon, allowing accurate predictions of the heavenly bodies as the Greeks understood them to be.

In studying the device with 3-D X-ray tomography and other innovative techniques, and by reconstructing its gear trains, scientists have discovered that it took into account both the Saros cycle (a period of eighteen years, eleven days, and eight hours that allows accurate eclipse prediction) and the Metonic cycle (a nineteen-year period for reconciling lunar months and solar years). The latter cycle is still used by NASA to calculate launch windows and by some Christian churches to fix the date for Easter each year. One of the dials of the mechanism also kept track of the four-year cycle of the Olympic and other pan-Hellenic Great Games.

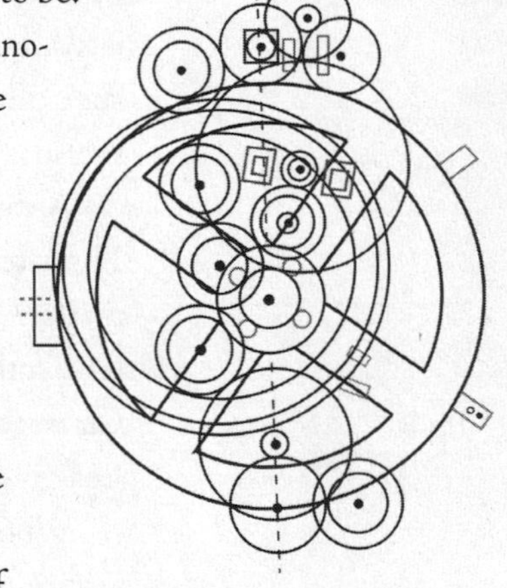

The Mac of its millennium, this early wonder calculated eclipses and other celestial goings-on—and came with user-friendly instructions.

Some speculate that the device sprang from the mind of Hipparchus, a nova-bright thinker of the second century B.C. This Carl Sagan of long ago closely tracked the movements of the moon over time, realizing that the darn thing had an elliptical orbit, not a circular one. That knowledge let him forecast the dates of lunar eclipses.

Such a breakthrough may seem underwhelming to modern minds, but in his day it was shocking news. Why? Because the Greco-Roman world ran on an intricate schedule meshing lunar and solar calendars. Festivals, farm work, and religious observances were lunar-dependent. Surprise eclipses? Rulers and generals hated them. Bad omens. Liable to cause panic.

Could the Antikythera Mechanism be an invention of Hipparchus, or one made to his specifications? The ship on which it was found carried items from Rhodes and its age correlates to his era.

Another strong candidate is Archimedes, the math and mechanical genius from Syracuse, Sicily. Inside the mechanism, researchers have identified the names of calendar months that match those of Syracuse and its Greek mother city, Corinth. The Roman author Cicero also described (and probably saw) the Sphere of Archimedes, a mechanical planetarium able to calculate the motions of the moon and planets.

A third candidate might be Posidonius, who later headed the school Hipparchus was said to have founded on Rhodes. Cicero was one of his students. According to surviving letters from Cicero, a device resembling the Antikythera Mechanism was invented by Posidonius. (Thank heaven for Cicero; the writings of this busybody, letter-writing fanatic, and astronomy freak provide a spectacular view into ancient life.)

Whoever dreamed up this precise device went much further than dreaming it up. He made it user-friendly. Inside this ingenious mechanism are imprinted detailed instructions for its use. Scientists are still studying the device, whose mysteries and functions continue to yield surprises. And they continue to look for its predecessors and successors. As Yale science professor Derek Price noted during his lifelong study of the device, as fully realized as it was, the mechanism must have been part of a long line of developments.

Using the built-in instruction manual, researchers from the multinational Antikythera Mechanism Research Project have built a working model. At their online site, visitors will ultimately be able to go back to 100 B.C. and calculate eclipses, using a three-dimensional model of the world's first protocomputer.

WHICH WAY TO THE PEARLY GATES?

Ancient ideas about heaven came from five sources, four of them Greek: the poems of Pindar and Hesiod and certain works of Homer, Vergil, and Plato.

The weirdest had to be Plato's fantasy called the "Myth of Er," which was tacked onto his book, *The Republic*. In it, a man gets a round-trip ride to the afterlife and learns about reincarnation, rewards for the moral, and ghastly punishments for tyrants, murderers, and certain other criminals. Oddly enough, the metaphor used is a lottery system run by the Fates; newly dead souls get lottery tokens to choose their next life. Plato envisions heaven as a fluffy sky realm where clean souls lived.

Most Greeks, however, didn't bother to look skyward for their version of heaven, since the poet Hesiod had said that the Isles of the Blessed were to the west, toward the setting sun and by the circling stream called Oceanus. Homer agreed on the location, calling his heaven the Elysian Fields.

Vergil, who lived after the aforementioned Greeks and became the favorite poet of first-century-B.C. Italy, sided with Homer but planted his Elysian Fields in a subdivision of Hades' underworld.

There I was, new scroll in hand, anxious to explain the Myth of Er, only no one showed at my booksigning.

If you MapQuested Vergil's version, you'd get the following directions: cross one of the five rivers of the underworld, look out for the mean three-headed dog on the opposite bank, go 1.4 miles through the fields of mourning, and turn right at the fork in the road. (Avoid left turns unless you want to end up in Tartarus or Erebus, Hades' hellish domains.)

The Elysian Fields boasted perpetual spring weather and lush groves of shade trees. Wearing snowy white garlands on their heads, the shades of the dead partook of an array of amenities, including sports. Shades could wrestle, for instance, on beds of clean yellow sand. No one ever got hurt or sick here, so no holds, wrestling or otherwise, were barred.

At the Elysian Fields, music appreciation and dance took place under the direction of the famed bard Orpheus. The newly dead were free to mingle with an elite crowd of heroes, patriots, virtuous poets, and priests who'd minded their celibacy vows. No streets paved with gold; it was a pastoral, sleeping-bag-under-the-stars sort of paradise. Of course, since it was underground, the Elysian Fields had to have its very own sun and stars.

Not everyone ended up there. There was also a down-at-heel place called the Asphodel Fields, named for the tasty little asphodel flowers of its ghostly meadows, described as the favorite food of the dead. The spirits of common ordinary Greeks, neither saints nor sinners, were shuffled off to Asphodel, where nothing bad happened, but nothing good, either. Since it was near Lethe, the river of oblivion and forgetfulness, residents were encouraged to drink from it. A bit of social snobbism operating in the afterlife, it appears.

Unlike the ebullient and vivid personalities they gave their gods and goddesses, Greeks and Romans didn't expend much time or imagination on heaven and the afterlife. People had ho-hum attitudes, unafraid but not overjoyed at the thought of an eternal vacation in the Elysian Fields.

When Christianity began to make headway into the religious mainstream in the second and third centuries A.D., its primary movers and writers loved the Elysian Fields concept, soon adopting it as a synonym for paradise, the better world awaiting after death.

They also cribbed from Plato as far as where heaven should be located. In their view, Christian sinners took the diamond lane straight down to the torments of hell, while the virtuous ascended to a sky heaven of surpassing beauty. Borrowing a bit from the Asphodel idea, they called the uppermost region of hell "limbo," defining it as a place where virtuous pagans (such as Plato) had a nice view of Christian heaven but could never get there from here.

These days, the Elysian Fields have gone on to a new afterlife, this time as the moniker given to countless less-than-heavenly housing tracts across the United States.

DUELING DOMAINS OF HADES

From earliest pagan times, various places in Greece and Italy have claimed to be *the* gateway to the infernal regions. The most popular site? The Phlegraean Fields. Besides being hard to spell, it sat in a foreboding, fire-and-brimstone sort of landscape, located in volcanically active Campania in southern Italy. Intrepid visitors were directed to the actual Mouth of Hell, found at Cumae near beautiful Lake Avernus.

Once there, they made an overnight stay at Cumae's Oracle of the Dead. This subterranean infernal theme park, carved from solid rock and terminating at hot springs underground, was in business to help inquirers make contact with an expired loved one. Details are understandably obscure; the visitor was kept in a sulfurous semidarkness to reach the proper degree of horrification, followed by a funereal meal of beans and milk and an entrail reading, then taken deeper to a sanctuary where he or she would hope to get a high-speed connection with the dearly departed.

Another place vying for most hellacious: Ephyra, in northwestern Greece, a hilly site that boasted two of the five rivers of the underworld. This venerable rival had a number of A-list clients by 600 B.C., including the Greek world's most notorious tyrant of the day, Periander.

A man given to perpetual road rage, Periander had kicked his wife Melissa to death because he'd heard gossip about her—from his concubines, no less. After the funeral, he couldn't resist giving his wife a farewell love poke, adding necrophilia to his murderous rap sheet. Later, he misplaced a household treasure and went to the Oracle of the Dead to ask Melissa a favor—not for forgiveness, however. Melissa helped Periander, clearly proving that the oracle's setup was a complete fraud, since no real-life wifely ghost would have let him off the hook.

Take a wrong turn in the afterlife, and you'd run into Cerberus the three-headed dog and perhaps Hades himself.

The decades-long archaeological excavations at Ephyra have uncovered massive structures and tunnels above and below ground, their function still debated. To the Greco-Roman way of thinking, the underworld covered a lot of acreage—so there was more than enough room for it to have multiple entrances.

The key thing about the ancient underworld had to do with hell, or more accurately, Hades. It wasn't a destination; Hades was a Greek god. The son of the deities Saturn and Rhea, he'd shown early promise, but his brothers Zeus and Poseidon got all the attention. To compensate, Hades cultivated a bad-boy reputation: kidnapping women, blackmailing men, becoming the selfish god of mineral wealth, and going by a variety of Latin and Greek aliases, such as Dis Pater, Orcus, and Pluto. Thus the underworld was Hades' domain—you can see how the confusion began.

Speaking of muddles, neither the Greeks nor the Romans agreed on what transpired after death. Some believed in the collective survival of souls, others in a form of reincarnation; still others asserted that there was no afterlife. Philosophers were equally torn.

Having absorbed various contradictory myths from childhood, the average Greek or Roman probably visualized the underworld as resembling one of our regional airport hubs. To enter, the newly dead paid a coin to Charon the old ferryman to take them and their luggage across the underground river called the Styx. At the equivalent of a ticketing counter, they would appear before Hades and two judges called Minos and Rhadamanthys, get sorted into "naughty" or "nice" categories, and get assigned an address in the underworld. There were subdivisions for heaven-bound spirits, too, called the Elysian Fields. Some souls ended up in a neutral standby zone; the soul also could try for an upgrade by making his or her way to Acherusian Lake to be purified.

Murderers and other obvious sinners in the bunch got assigned to Tartarus, aka Erebus, a gated community of hell where they spent their time wincing at big-time sufferers such as Tantalus and Sisyphus. Once in a while, a flesh-eating demon would show up to break the monotony. After a year of this, criminals who claimed extenuating circumstances or poor legal representation could go to the special lake mentioned earlier. There, if they told their (presumably dead) victims how incredibly sorry they were, they got to hang out at the shore with the others.

Except for Tartarus, the maximum-security section, Hades' home sounded dull but restful. There were no blazing fires, no screams of the damned, no satanic big boss man. Instead of horns and a tail, its CEO carried a staff to prod the newly dead downstairs and wore a cool helmet that conferred invisibility. (*Hades* actually means "invisible" or "unseen.") Instead, it was the unquiet spirits of the dead, rather than the place where one's soul might end up, that preoccupied the minds of the living.

No matter how scandalous the behavior of Auntie Julia or how cranky and flatulent Grandpa had been, once they passed on, they were considered *di manes*, the divine dead. Spirits liked to hang out close to home turf; moreover, they were terribly touchy, so it was important to appease them. For that reason, gravesites, mausoleums, and niches that held cremated remains had small pipes or holes running into the earth as standard equipment. At funerals and on special occasions throughout the year, family members carefully poured wine and not-too-chunky food items into the holes. Their object: to make sure that Granny's ghost stayed subterranean. To be extra prudent, survivors sometimes weighted down the bodies of their newly departed with lead—an expensive touch, but worth it for the peace of mind.

NO MATTER HOW WEIRD, JUST DO IT

Greeks were always trying to scope out what the gods meant, what this or that goddess liked or disliked. They also sought divine help with more quotidian matters: the outcome of a pending lawsuit, that chronic rash on Grandpa's buttocks, why a decent husband couldn't be found for an oldest daughter. In search of these truths, they frequently dipped into the rich cornucopia of divination methods available to them, such as oneiromancy or dream reading.

Although Epicurus the philosopher scoffed, saying dreams had no divine nature or prophetic force, his was a lonely voice. Most people preferred to believe Homer's words: "Dreams come from Zeus." Or Plato, who thought dreams were one of the ways that the gods conveyed their intentions to mankind.

To puzzle out the meaning of dreams, corps of interpreters were standing by on every street corner. For the do-it-yourselfer, guidebooks were readily available, such as *Oneirocritica,* the bestseller by Artemidorus of Ephesus in the second century A.D. (Talk about in your dreams—other authors can only daydream about producing such a gold mine, still in print in various formats nineteen centuries later!)

Some dreams required active follow-up on the part of the dreamer. The incubation dream, often taken by patients seeking cures for mysterious ailments, required overnight stays in holistic healing centers called *asclepeia,* which could be found across the Greco-Roman world. One famous "Just do it!" dream happened to Anyte, a Greek poet of Arcadia who specialized in lyrical epigrams. In her slumber one night, the healing god Asclepius handed her a set of sealed wax writing tablets with orders to deliver it to a certain Phalysius of Naupactus. When she awoke, she found the sealed tablets in her hands. Back in 300 B.C., you didn't take an omen like that lightly. Anyte packed an overnighter, then trudged on foot north through Arcadia to the coast, finally locating a sailing vessel bound for Naupactus—no piece of cake, either.

At length she ended up at the door of Phalysius, where a nearly blind man answered.

Anyte told him to break the wax seal and read the contents of the tablet, pronto. A little affronted at this travel-stained young woman who insisted she'd brought the equivalent of a text message from the god Asclepius, he struggled to make out the words on the tablet.

Not surprisingly, his eyes were instantly upgraded to 20/20 vision. Even better from Anyte's perspective, he followed the instructions on the tablet—and gave her 2,000 golden staters! Not a bad day's paranormal adventure.

It was a dream world triumph all around, because the newly keen-eyed Phalysius immediately built a sanctuary to the god Asclepius in the little burg of Naupactus. We know the whole story because Pausanias, that indefatigable travel writer and historian, researched the story on-site and wrote about it.

Warm and fuzzy tales of dreams come true aside, written guides to dream interpretation tended to run a bit amok in their explicitness. Then, as now, sex sold. Ancient write-ups contained quite a few entries along the lines of the following from *The Interpretation of Dreams* by Artemidorus: "To dream of having sex with the moon: auspicious for pilots, merchants, tourists, and tramps. For others, an attack of dropsy."

Thorough to the point of obsession, Artemidorus covered every quirk and subcategory of sex dream, including having it standing up, with children five and under, with one's living mother, and with a maternal corpse. This last necrophilia, he advised, was generally bad news dreamwise—unless you were involved in a lawsuit over land, of course.

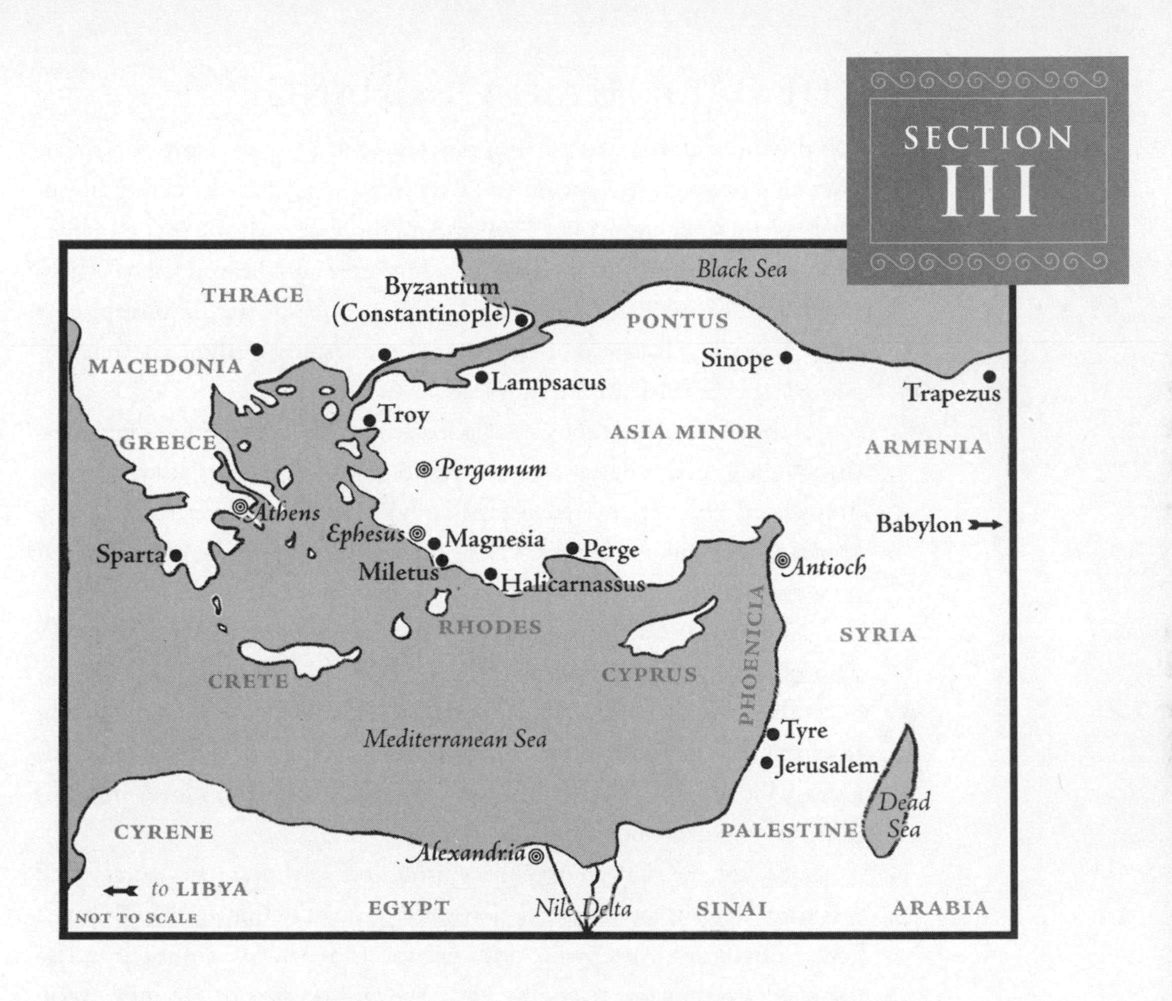

Asia Minor & the Middle East

THE DARK HEART OF CHANGE

His doctrines anticipated modern physics—but he spent most of his time offending everyone in sight. They called him the riddler, the dark philosopher, the melancholic man of physics. And those were the nicer comments.

In the sixth century B.C., when Heraclitus came into the world, the Persians ran his corner of it and had for half a century. Ephesus was his birthplace, a florid Greek city balanced on the coast of the eastern Mediterranean countryside, a part of Asia Minor now known as Turkey.

As the oldest son of the city's figurehead ruler, he could have inherited the mantle of ersatz power with few responsibilities. Politics, however, gave him a headache. He sneered at his family's pretensions, preferring to play knucklebones with the gang that hung out at the Artemisium, Ephesus' great temple to the goddess Artemis. But soon that got old, too. Heraclitus was a precocious thinker, an intellectual who "inquired of himself," a gloomy fellow who let fly at targets big and small.

Unlike most philosophers and natural scientists, he had no mentor or teacher. Contemptuous of others, rigidly ethical, and way too brainy for the average Greek, this prickly Ephesian took care to couch his ideas in terms difficult to understand.

Until he came along, most elite eggheads believed that the natural world rested on an unchangeable foundation, ruled through eternity by the Olympian gods. Heraclitus turned their worldview upside down. "We both step and do not step into the same rivers," he said. "We are and are not." To him, everything was in a constant state of flux or flow—with an underlying order to this change. He also believed in the unity of opposites, calling it the upward-downward path. No day without night, no good without bad, no summer without winter. This man saw a cosmic balance in the struggle between fire, air, water, and earth, the elements from which he thought all was made. Fire, the transforming power. Fire, the expression of the deity in everything. Fire, which seemed to destroy but only transmuted. In that sense, Heraclitus was a chemist as well, perceiving that fire recombined elements, but their atoms still remained in existence. These concepts blew the doors off most of the other philosophers of the Milesian and Ionian schools, who postulated that objects either existed fully or did not exist at all.

Heraclitus wrote down his doctrines in a work called *On Nature*, which he placed at the temple of Artemis, his youthful hangout. That sole copy was divided into three sections on politics, theology, and the universe. He refused to have the work published or copies made, saying that those who truly sought the truth would find it. And they did, for centuries. Some were irate; Plato had a hissy fit, sputtering, "How can that be a real thing which is never in the same state?"

Despite his deplorable bedside manner, Heraclitus did win converts. Simplicius wrote about him later, recasting one of his main premises as *panta rei*—everything flows, nothing abides. Philosophers from Theophrastus to the later Stoics, from Cleanthes to Roman emperor Marcus Aurelius, integrated Heraclitus' landmark insights into their own theories.

Everything flows, nothing abides. Except for my acid reflux.

If anything, his flame burns even brighter today. Much of what Heraclitus put forth has found modern parallels and proofs in physics, math, and metaphysics. He's also looked at as an early pantheist, denying the existence of the gods while proposing instead a living cosmos, a grander unity.

As he got older, Heraclitus became more firmly misanthropic. His blunt words and manner made far more enemies than friends, although he did have a buddy named Hermodorus, to whose defense he once sprang. He growled, "The Ephesians would do well to hang themselves, every grown man of them, and leave the city to the beardless lads, for they have cast out Hermodorus, the best man among them."

At length he abandoned city life altogether, and took to the hills, trying to live on a diet of herbs and greens—never an easy task for a bilious personality. At a certain point, now nearing sixty, he came down with a ruinous case of dropsy, an edema that can surround the heart or kidneys and lead to diabetes. Heraclitus unwillingly sought medical treatment. To inform the doctors about his condition, he asked them, "Are you able to create a drought after heavy rains? Can you empty my intestines by drawing off the moisture?" His disease, or perhaps his enigmatic words, baffled the medicos.

Heraclitus stormed out of the place and sought his own cure. Finding a large pile of warm cow manure, he buried himself in it, to "draw out the noxious humours," as he put it. Reports on his demise differ, but one has it that as the cowpies dried, Heraclitus attracted unwelcome attention—and not all of it from his fellow men. The town dogs of Ephesus found him irresistible. Heraclitus died in dung, devoured by unfastidious canines.

Although he said that everything flows, nothing abides, in his case that's not precisely true. What Heraclitus thought, what he taught, has abided, flowing into a greater stream.

TORCH SONG FOR MARBLE MANSIONS

The Greeks had an edifice complex. So did the Romans. They built palaces, government offices, and temples to honor an embarrassment of deities. As history reveals, however, what went up came down with surprising frequency. From the Pantheon to the Parthenon, civic structures were destroyed by fires, earthquakes, and human barbarism. What's more, most were rebuilt, only to be burned down or wiped out a second or third time.

Okay, earthquakes we can grasp. Or demolition by invading armies. But how could the marble pillars and stone arches of temples and amphitheaters

The Artemisium topped the World Wonders list but fame couldn't save it from firebugs. Read why.

go up in smoke? To unravel the mystery, a closer look at one architectural triumph is in order: the Artemisium at Ephesus in Asia Minor, an official world wonder for nearly ten centuries. This sacred house, lavishly underwritten by Croesus (yes, that Mr. Rich as Croesus), enabled an architect called Chersiphron to embark on a marathon project.

Because the region was earthquake-prone, the architect deliberately built on marshy estuary land. To begin, he had workers dig deep trenches and pour masonry, then put down layers of crushed charcoal, alternating with layers of sheepskins, their wool unshorn. This provided the temple with a springy underpinning that "gave" with seismic activity.

When finished, the Artemisium was a showstopper. Located at the intersection of Europe and Asia, it shimmered with the color and drama of the East, coupled with a Greek sense of harmony. Once visitors climbed the twelve broad steps of the platform, they saw a classic rectangle in the lush Ionian Greek style. They walked among a noble forest of 127 ornately carved columns, each commissioned by a different royal big shot. The 60-foot columns circled all sides of a building 425 by 225 feet—the interior occupying over two acres.

To reduce weight on the structure, the roof had an open area cut into the center. That let more light into the inner chamber where the cult figure stood: a mother goddess, Lady of the Animals, much older than Artemis of the Greeks (or the Roman version, called Diana). Her cult statue had what looked like abundant rows of female breasts sans nipples; historians now believe they represented gourds, testicles, or other symbols of fertility.

Greek and Roman temples weren't designed to hold worship services. Activities such as sacrificing took place outside, at altars in the sacred precinct. Besides housing the cult figure of the deity, temples became storehouses for votive offerings. Over time, these collections grew mountainous: from gold statues, bronzes, and jewelry to the B.C. equivalent of macramé items, kitschy wall hangings, ex-voto body parts, and other junk.

Since sacred houses also functioned as banks, below its main floor the Artemisium had safe deposit vaults, where the wealthy stored their loot for safekeeping. At this temple, there would have been leather sacks filled with the first gold coins ever made, each stamped with the image of the ancient goddess.

Besides piety, tourism, and banking, visitors had other reasons to come. The art gallery contained works by Apelles and other big names; the place also gained fame for its choral musical offerings. To cope with visitor traffic, the Artemisium employed a staff of maintenance folks, security men, clerks for money transactions, guides for tours of the interior, plus entrail readers, priestesses (called *melissae*, "bees"), and other helpers. About the only amenities not on-site were restaurants and sanitary facilities—but the city of Ephesus had those aplenty.

Most temples were open night and day. As a result, they depended heavily on artificial lighting and heating via oil lamps, torches, candles, and charcoal-burning braziers. To stoke them, ample supplies of olive oil and firewood were stored on-site. Open flames burned continuously on the sacrificial altars outside the building as well.

Crowned with ceramic roof tiles, the Artemisium sported glorious expanses of marble, along with marble columns and floors. Like every Greek temple, however, the framing of the roof and walls, the main beams and the architraves positioned on top of the columns, the doors, window frames, and other fixtures were made of wood. Dry, seasoned wood.

After a temple had been in business a few years, the sheer quantity of offerings turned interiors into a landscape of flammable stuff. Even the marble columns got swathed in tinsel, military flags, and draperies of wool and silk.

There were other fateful elements that led to the burning of the Artemisium and other grand buildings. The first involved the workings of nature and that attribute of Zeus, the lightning bolt. Temples and other structures had to withstand as many storms as we experience today, but largely without the protection of lightning rods. They also lacked adequate firefighting equipment and technology. Even in later centuries, with Roman hydrological expertise, there was insufficient water pressure to fight conflagrations. High-volume hoses were rare commodities as well.

Temples such as the Artemisium were built to be seen. They usually hogged the highest ground—promontories, headlands, or hills, all places where lightning repeatedly strikes. (These days, cell phone towers and the Empire State Building receive thousands of hits each year.)

As a lightning bolt strikes, it zings to a peak temperature five times hotter than the surface temperature of our sun—a tad more than needed to set anything aflame. Once a fire reaches 1,000 degrees Fahrenheit, nonflammable materials such as marble, clay tiles, and limestone start releasing carbon dioxide and turn to powder.

Another piece of the puzzle is the tragic but deeply human strain that runs through history: the pathological glory hound who seeks fame through destruction and death. In the case of the Artemisium, a young man was responsible for its fiery downfall in 356 B.C. When caught, he readily confessed. Even bragged. The Ephesians were so incensed, they executed him—and inflicted the death penalty on anyone who ever mentioned him by name again. (And I'm not about to tell you, either.)

Only then did they start building a bigger and better Artemisium, which stood firm for six hundred more years until those blasted Goths broke it into smithereens in A.D. 262.

LEVITATION CAN BE ELECTRIFYING

The Greeks never knew the delights of a power grid, but they certainly recognized the power of things electric, from lightning to electric eels. Slimy specimens of the latter were employed in Greco-Roman times to ease the pain of gout attacks. Talk about chills and thrills: sufferers placed their feet on enraged captive eels until their gouty digits went numb.

The word *electricity* derives from *electron*, the Greek word for "amber," one of several substances that will hold a small electrical charge when friction is applied. Amber itself, millions of years in the making, is the fossilized resin of several extinct species of conifers. More than twenty-five hundred years ago, a sage named Thales noticed amber's ability to attract lightweight objects such as straw.

Another Greek naturalist named Theophrastus regularly wandered the countryside in search of interesting plants. While doing so, he stumbled on a stone he called *lyngourion* (tourmaline, in all likelihood), which had the same power to attract small pieces of matter. Fiddling with it, he felt the tingle of static electricity. These discoveries, while fascinating, were useless for generating energy but did generate hundreds of imaginative superstitions.

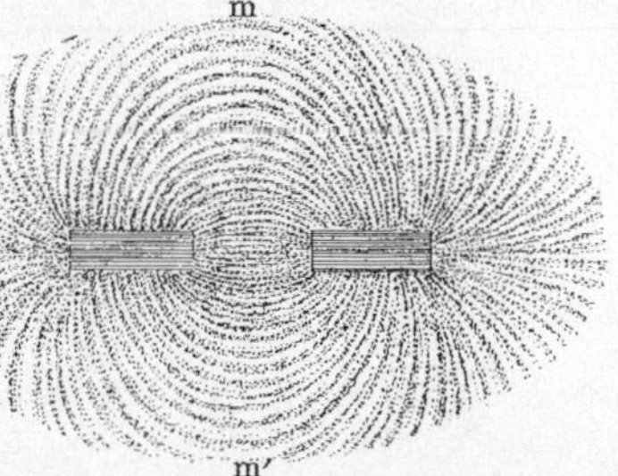

About the same time as his amber find, Thales got excited about lodestones, the naturally occurring magnets known nowadays as magnetite. As it happened, Thales lived in beautiful downtown Miletus, a superbly harbored city-state on the

Maeander River, a saunter away from another splendid city called Magnesia on the Maeander. The landscape of Magnesia happened to be loaded with magnetic iron ore, lousy with lodestones. After Thales' encounter, the local chamber of commerce—knowing how the media loved a good story—circulated a fairy tale about a shepherd wearing shoes with iron nails who found his feet clinging fast to Magnesia's rocky soil. No one bought it, since the idea of shepherds wearing shoes was laughable. Still, the place name Magnesia did stick magnetically, as it were, to magnets.

Initially, the main use for lodestones was medical. Doctors applied lodestone powder on burns and as eye salve. Larger pieces may have been used to extract splinters from eyes or throats.

Meanwhile, in parts of the globe quite unknown to the Greeks, the Olmecs of Central America were using magnetite to orient their temples north-south. In a different quarter, the Chinese employed lodestones to make divining boards, later seeing their value as magnets to make the first decent compasses. In Egypt, lodestones, called "north-south iron" or "bones of Osiris," were old hat. Being very fussy about pyramid alignment, the Egyptians had found the polarity of lodestones extremely useful.

Back in Greece, however, the magical qualities of these attractive minerals and resins mainly served to enliven dinner parties and increase the numbers of awestruck visitors to Greek oracles and religious shrines.

How did lodestones work, anyway? It baffled the brain, but various theorists had a crack at it. Diogenes of Apollonia (no, not that Diogenes) announced that iron contained lots of humidity, attracting the parched, dry magnet, which fed upon it. (Greek thinkers were big on dryness/wetness and other opposites.)

Unlike today, where magnets are the unsung workhorses on our refrigerators, holding up a universe of curling photos and memorabilia, in B.C. times people got a terrific kick out of playing with these inexplicable objects.

About 430 B.C., Socrates was quoted as saying that lodestone "not only attracts iron rings, but imparts to them a similar power of attracting other rings; and sometimes you may see pieces of iron and rings suspended from one another to form quite a long chain; and all of them derive their power of suspension from the original stone."

Early Roman scientists such as Lucretius marveled at their uncanny ways. "It also happens sometimes that the nature of iron withdraws from this stone, and will flee it and follow it in turn. I have even seen iron rings leap up, and at the same time iron filings seethe in bronze bowls when a magnet has been put underneath, so eager to escape the stone is iron seen to be."

In the third century B.C., another marvel came along—a woman who possessed animal magnetism, a force at times more powerful than lodestones. For her third nuptial, Arsinoe II opted for true love, tapping her brother Ptolemy II. This was Alexandria, Egypt, where sibling rivalry alternated with sibling revelry. After her quick wit got her easygoing brother out of various political jams, he renamed whole sections of Egypt for her, and had Arsinoe officially declared a goddess.

Ptolemy then hired Egypt's top architect to begin work on an elaborate temple to his sister-wife. It would feature her glittering statue, suspended goddess-like in midair, the whole apparatus using large lodestones. Arsinoe was beside herself with joy. Maybe excessive joy, since on a sweltering July day in 270 B.C., Arsinoe was quite chagrined to expire of what was called natural causes at age forty-six. Postmortem, her brother-husband showered more honors on her. Unfortunately, both he and his architect died before the statue of his high-flying adored could be completed.

Despite these mortality setbacks, the technological quest for using magnetism to suspend objects refused to die. Yet another Ptolemy built the Temple of Serapis in Alexandria, its new world wonder an iron chariot said to be held in midair by the powers of the god. It drew admiration for decades.

Other temples in Egypt and Gaul copycatted with magnet-enabled devices of their own. The poet Claudian described the charm of an Alexandrian one. Its centerpiece held two statues: an iron figure of the warrior god Mars and a beautiful Venus carved from a large lodestone. At celebrations open to the public, the two figures were placed near each other on a reclining couch covered with roses. As music played (and with a bit of stage management), the iron figure of Mars finally "leaped" unassisted into his lover's arms.

Ancient magnets and gleaming nuggets of amber and tourmaline always had a glamorous aura, a playful air about them. Not so the always sobering phenomenon of lightning. Early in Greek history, thunderstorms had thoughtful seekers searching for something besides incantations and amulets to protect against the devastating power of lightning strikes. The quest may have begun in Minoan times. One clue: archaeologists have found what plausibly appear to be lightning rods that once stood sentinel on the roofs of small temples in Minoan Crete, especially ones sited on mountain peaks. In Egypt, researchers have also encountered copper-sheathed lightning rods attached to the tall pylons or gateways to Egyptian temples, inscriptions about which still can be read on the buildings. These Greco-Roman artifacts date to the third century B.C. If these indeed were lightning rods, they never caught on in a bigger way, or perhaps we have yet to find key evidence.

Although inquiring minds such as Lucretius and others in that long-ago world recognized the parallels between the electrical attraction of amber, the power of lightning, and the magnetic forces of lodestones, they didn't understand how close that relationship was. Only in our time has the physics of the electromagnetic field become accepted as one of science's fundamental forces, making possible everything from power grids to maglev bullet trains.

AN UP-AND-DOWN PROPOSITION

More than once, meteors made Greeks famous. Not only that, but meteorites that survived the fiery trip to earth often became cult objects of worship—a devotion that continues today, believe it or not.

The story properly begins with Anaxagoras, an Ionian Greek from the Asia Minor city-state of Clazomenae, who brought a spirit of scientific questioning, and much else, to Athens. Taking the universe as his focus, he developed an early version of the Big Bang theory and taught his brand of philosophy in the city. By midlife, he had blazed a path to right answers, wrong answers, and near misses about things celestial. Among his hits: lunar eclipses were caused by the earth's shadow on the moon. Among his misses: the earth was flat, the sun was a blazing stone bigger than the Peloponnesus, and there were dwellings on the moon. Anaxagoras also asserted that heavenly bodies were stones, made red-hot by their motion through the sky—and warned that only centrifugal force kept the stones from raining down on him and everyone else.

After intense study, in 467 B.C. Anaxagoras made a bold prediction: within a short time, one of those scary big red-hot stones would fall to earth. When a brown meteorite the size of an oxcart landed in broad daylight at Aegospotami, a riverside town hundreds of miles north of Athens, the philosopher won his bar bet and larger acclaim among the Greeks.

I really hate to part with my favorite meteorite but since you're Rome, and I'm not . . .

A couple of decades down the road, though, the poor guy was accused of impiety—the "failing to recognize the sun as a god" statute—and brought to trial. Since this was a time of anti-intellectualism in Athens, it's likely the charge was politically motivated, especially since Anaxagoras was close friends with Pericles, the former top dog, who'd fallen out of favor. Because impiety carried the death penalty, this celestial whiz dodged the meteor fallout by leaving to set up a teaching school in Lampsacus, back among the more scientifically tolerant Ionians.

Meteor prediction as a career move took a nosedive after that, but the objects themselves continued to burn through the atmosphere, some landing intact. (No one yet had a clue that such meteors are hitchhikers, the long-lived entourage of comets. Or that each May, we see meteor showers that still may contain ancient particles shed by the passage of Halley's Comet during the time of Julius Caesar.)

In the third century B.C., a handsome specimen crashed down near the Asia Minor city of Pergamum. Brought to the attention of King Attalus I, Rome's only ally in the region at the time, the shiny black meteorite became the centerpiece for local worship of the mother goddess.

A few years later, around 205 B.C., leaders of the Roman Senate began to stress about the spate of meteor showers over their city. After consulting their revered Sibylline Books, it became clear they needed help from a mother goddess, a commodity they were fresh out of in Rome. Someone remembered Attalus; a delegation was sent. The king, eager to please the up-and-coming Roman superpower, regifted his meteorite. It was duly installed in a brand-new temple on Palatine Hill and dubbed Magna Mater—the Great Mother.

MM was still there four hundred years later, when Rome got stuck with a particularly embarrassing emperor from Syria (by then part of the Roman Empire). Calling himself Elagabalus, this teenage devotee of the sun god absolutely refused to make do with someone else's sacred black stone. As the cross-dressing high priest of his own cult, he had his own holy roller hauled all the way from Nicomedia to Rome and housed in a far more ostentatious new temple.

The Romans were not easily rattled. With a sigh, they accepted the boy's attempts to make meteorite worship the only game in town. They tried to overlook his sacrilegious activities, such as his marriage to a horrified vestal virgin. They snickered at Elagabalus' attempts to expand his sexual horizons by asking his doctors to make him an artificial vagina.

But the emperor finally crossed the line when he neglected the Roman army and attempted to rub out his cousin Alex, his heir. In short order, the disturbed young head of state ended up as a mutilated carcass in the Tiber River.

And Emperor Elagabalus' sacred black stone? It was returned with considerably less ceremony to Syria, where the large cone-shaped meteorite continued to be worshiped by locals.

Similar sky rocks also caught the fancy of religious cults in other locales; the island of Cyprus had two. Later, when another shiny black meteorite was found on the Arabian peninsula, it became the cornerstone of the Ka'aba in Mecca, still the holiest spot in the Muslim world.

NOT EASILY ECLIPSED

Hipparchus wasn't the first Greek to forecast eclipses. At the forty-eighth celebration of the ancient Olympic Games, in front of a crowd of fans, a confident intellectual named Thales announced that a solar eclipse would occur on May 28 of 585 B.C. He was right on target—and his prediction is called the first exact date to be confirmed in ancient history, winning Thales a berth on the Top Seven Sages of Greece list.

In Thessaly, a remote region near the Olympus Mountains, tradition had it that female sages with mysterious powers were able to "call down the moon." The legend probably rested on the real-life career of Aglaonice, a stargazing Thessalian who regularly predicted lunar and solar eclipses. She was said to have deep knowledge of the Saros cycle, an eighteen-plus-year period discovered by the ancient Chaldeans of Mesopotamia.

Centuries after Thales and Aglaonice, Hipparchus was born in 175 B.C. As he grew to manhood in Asia Minor, his intellectual feats soon put other astronomers in the shade. Instead of predicting a one-off, Hipparchus laid out a years-long schedule of lunar and solar eclipses in his part of the Mediterranean. In doing so, he got rid of some of the dread that people felt about eclipses.

Astrolabes—what a hassle. It'd be a lot simpler if I just plagiarized Hipparchus' results.

To the average Greek in the street, comets were dire, but eclipses screamed disaster. The frightening way in which the sun or moon got swallowed up by some shadowy force—surely that suggested something terrible was about to occur on earth. The unexpected death of a leader, perhaps, or a plague or similar tragedy. Athenians still winced over their city-state's ghastly defeat in the Sicily expedition of 413 B.C., when a lunar eclipse caused their forces to stay put.

In those days, both the sun and the moon were routinely called stars or wandering planets. Since they were personified and worshiped as the sun god and moon goddess, an eclipse was also looked on as an attack on their heavenly persons. A lunar eclipse was seen as the moon goddess dying; to come to her aid, musicians would gather in ad hoc groups, playing the cymbals as loudly as they could.

To get the best naked-eye visibility for his work, Hipparchus moved to the island of Rhodes in the Aegean Sea. There he spent the rest of his days, carrying out his stellar work. He got in on a lot of eclipse sightings. By the sound of it, the most spectacular occurred on November 26 in 139 B.C. Looking out to sea as the sun was rising, the astronomer turned to catch the uncanny sight of the moon being eclipsed nearly simultaneously in the opposite direction.

Deep into research, Hipparchus discovered that eclipses ran in regular cycles, with lunar ones occurring only at the full moon, and solar ones falling when the moon was in its first or last phase. He also showed that eclipses in the Northern Hemisphere were not necessarily visible in the Southern Hemisphere, and vice versa.

This talented astronomer set himself a number of celestial questions to answer. He drew heavily on earlier work, such as the Babylonian method of dividing a circle into 360 degrees, or 60 arc minutes. To learn more about ratios, he invented trigonometry, the science that measures angles in order to find the unknown sides of triangles. He also invented the astrolabe, a device to find the geographical latitude by observing the stars.

The most valuable discovery he made? Precession. Not an easy concept, precession—especially in his time, where red-hot debates raged among astronomers about just what was in the heavens, and what part the earth played. Many adherents clung to the everything-moves-around-a-static-earth idea. A few loony trailblazers believed that the earth moved around the sun.

Hipparchus set out to prove an even wilder notion: that the earth itself revolved, and, like a top winding down, wobbled in its orbit.

He ran across this puzzle while examining the ways people had of measuring the length of the year. None of them matched up. Among other dilemmas, the tropical year (the time it took the sun to go from one spring equinox to the next) was about 20 minutes shorter than the sidereal year (the time it takes the earth to orbit the sun). To solve his celestial enigma, Hipparchus took the longitude of Spica and other bright stars, then compared them to measurements made by earlier colleagues. He was able to prove that the equinoxes were slowly precessing (a fancy way to say moving) through the zodiac of constellations in the night sky.

Although it's possible that the astronomer may have thought the earth did not have an orbit of its own and that the twinkling heavens slowly revolved above the earth, his insight about precession still holds true. And his calculation of 25,771 years for the cycle to complete itself—called the precession of the equinoxes—remains valid.

Eager to learn more, Hipparchus studied the work done fifty years prior by Aristarchus of Samos. Applying the methods of his fellow astronomer, he pegged the size of the moon to its actual one-fourth-of-the-Earth diameter. Hipparchus then took a swing at the earth-to-moon distance, getting an "in the ballpark" result. (Although often given as an average of 240,000 miles, the moon moves as close as 225,740 and as far away as 251,970 miles from the earth.)

As his final contribution to science, a now-elderly Hipparchus compiled a star catalog, using the armillary sphere he invented to model the heavens. He mapped the positions of 1,080 stars, including their degree of magnitude or brightness.

Besides the star catalog, Hipparchus wrote a number of books on astronomy. Only fragments survived, picked up in the work of others—principally by a later scientist who worked with far less accuracy, to say nothing of scruples. That man was Claudius Ptolemy of Alexandria. Ptolemy not only lifted all of Hipparchus' star sightings for his own catalog, *Almagest*, but actually got the data wrong as he transposed the figures to his later century.

Looking on the bright side: since the bulk of Hipparchus' original work has vanished, we can thank *our* lucky stars that plagiarizer Ptolemy cribbed so much from Hipparchus.

GIRLY MEN AND CONTACT HIGHS

Herodotus, invariably called the father of history and less flatteringly, the father of lies, was a brilliant innovator, the first to attempt the epic scope of history through more than battles and kings. With the help of concepts such as kinship and reciprocity, he sought to explain the conflicts between nations and individuals by studying their customs, religious beliefs, myths, physical surroundings, gender relations, and antecedents. In short, he became the Margaret Mead of ancient times—an eager anthropologist, at times naive, at others exceedingly astute.

Scythian warriors loved to sew. Short on fabric but long on macho, they used human hides instead.

Fortunately for us, he often let his curiosity get the better of him. The stories he collected (some of which he pooh-poohed but allowed to stand because they were so darned juicy) and the digressions he included are eye-popping. In his work, called *The Histories*, many aren't digressions at all but background for upcoming sections of the book.

Being from Halicarnassus in Asia Minor (modern Bodrum, Turkey), Herodotus was very familiar with the Greek city-states, the Persians (who controlled his region in his day), and other non-Greek cultures around the Black Sea. One society that fascinated the writer-historian was that of the Scythians, the nomadic über-warrior culture of horsemen that ran from the north shores of the Black Sea deep into the regions we now call Russia and the Ukraine.

The Scythians' bloodthirsty ways made Greek warfare look tame. Teenage boys had to make their bones by slaughtering an enemy, drinking his blood, then cutting off his head and making a scalp hankie out of the flayed skin. Adult males got very elaborate with their flayed-skin needlework, sometimes making arrow quivers out of human hands, being careful to include the fingernails for the best fashion statement.

A frugal bunch, the warriors also sawed off the skull tops of the defeated to make drinking cups—some of which they upholstered with oxhide and gilded. Naturally, when drinking from said cups, the men had to swear terrible oaths, which required them to punch a hole in their bodies with an awl or whatever was handy, mix the blood with wine, then stir with weaponry before drinking.

This all sounds very artsy-craftsy macho, but the Scythians had their softer side—a designated group of male soothsayers they called Enarees, meaning men-women. While examining their culture, Herodotus learned that on a long-ago raid, Scythians had blundered by plundering a temple of Aphrodite (the earlier Greek version of Venus). This inflamed the goddess, who imposed a permanent curse of impotence and/or hermaphroditism on certain male descendants.

Even amidst the blood and testosterone, the Enarees had a fairly relaxed time of it—except when it came to actual prophesying. This was done in a competitive, reality-show atmosphere. Whenever the Scythian king came down with the flu or acid reflux, three soothsayers had to finger the guilty party. Should the alleged cause protest his innocence, six more soothsayers were brought in. If they condemned the man, he was beheaded and the first fellows divvied his belongings. If however, they disagreed, more soothsayers were called. And should a quorum happen to acquit, the first three ended up on a red-hot bonfire.

But those tense moments weren't the Scythians' sole idea of fun. When not on horseback, they hung out in tents and chilled as much as any civilized, non-blood-drinking people.

It so happened that on their rich green lands another sort of grass grew in abundance. Instead of smoking it, the Scythians pitched an auxiliary tent and lit a charcoal fire inside, using a small hibachi filled with stones. When it reached a toasty heat, they threw cannabis seeds onto the blazing stones and proceeded to get stoned.

Herodotus, who writes like he's experiencing a contact high, described them as "howling, awed and elated by the vapor." Their method combined the psychoactive qualities of cannabis with the cozy ambience of a steam bath—and chances are good that no one ever bogarted that joint. With a slight whiff of distaste, Herodotus added that Scythian males never washed their bodies. Instead, they regarded their cannabis sessions as "bathing."

Scythian women, however, went a more meticulous route. Grinding together a mixture of cypress, cedar, and the woody stems of frankincense, they covered their faces and bodies with a thick plaster of the stuff. The next day, they cracked off the dried plaster and as Herodotus put it, "emerged sparkling clean." The whole cannabis-hygiene routine has the ring of first-hand reporting.

This portion of Herodotus' book used to provoke snorts of incredulous laughter from historians—until archaeologists in the twentieth century started finding Scythian stashes that contained get-high kits of tent poles, braziers, and scorching vessels, along with caches of hashish and opium in other nomadic regions adjacent to the Scythian steppe.

As a result, researchers now tend to view his more outré statements with a kinder eye. Herodotus loved a good story, but that "father of lies" label was triumphantly off-base. Like our own world, the truth of his world was stranger than we can imagine.

GOLD-GUARDING GRIFFINS TO DINOSAURS

Among the fabled beasties and monstrous beings dreamed up by the imaginative Greeks, the griffin stands out on three accounts. First, its existence was reported on by the likes of Ctesias, Aelian, Herodotus, Pliny, and Pausanias and described in essentially the same way for a thousand years. Second, all sources agreed that griffins were to be found to the east of the vast lands roamed by Scythian nomads. Lastly, the critters were always associated with gold—sometimes guarding it, other times collecting it in their nests.

What did your typical griffin look like? As Ctesias put it, "They were a race of four-footed birds, almost as large as wolves and with legs and claws like lions." Fierce but flightless, the griffin had a huge cruel beak and made nests in the foothills of desert mountains.

Along with others, the Roman author Aelian wrote about the struggle between those who sought gold in faraway regions and the griffins that supposedly guarded it. As he said, "Dreading the strength of griffins, the [nomadic Scythian] miners avoid hunting for gold in the day. They approach at night when they are less likely to be detected. The place where the griffins live and the gold is found is a grim and terrible desert. Waiting for a moonless night, the treasure-seekers come with shovels and sacks to dig. If they manage to elude the griffins, the men reap a double reward, for they escape with their lives and bring home a cargo of gold—rich profit for the dangers they face."

The Greeks have always loved a good story, so it's easy to smile tolerantly at those ancient spinners of tall tales. Paleontologists and classical folklorist Adrienne Mayor, however, have discovered intriguing links to the griffin saga. What neither Aelian nor the Scythian gold-seekers knew was that the beasts they called griffins had lived in a much earlier era. The beastly bones that terrified them hadn't died recently. Au contraire; they were between 65 and 100 million years old.

Far to the east of the Caucasus and the Russian steppes where the Scythians held sway sit the brooding reddish sands of the Gobi and other remote Mongolian deserts. They are chock-full of white dinosaur bones, including rich pickings from more than one genus. The conditions are perfect for preservation, and many complete skeletons have been found

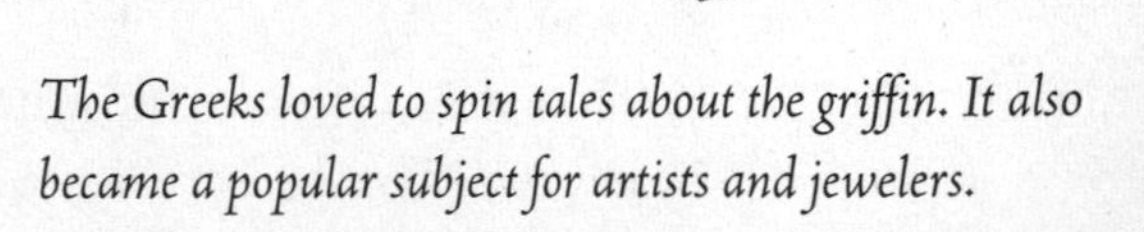

The Greeks loved to spin tales about the griffin. It also became a popular subject for artists and jewelers.

since the mid-1950s, including Citipati, Oviraptor, and Protoceratops. (Genus and species names are still in flux, due to nearly constant new finds in the area.) In general, these four-legged dinosaurs combined the traits of large flightless birds with mammal characteristics. They sported large beaks, heads with bony crests, and redoubtable claws, along with long and powerful tails.

As modern gold miners and scientists have found, the desert regions where these dinosaur species roamed do contain gold, much of it lying near the surface. It often washes down from the Altai Mountains as gold-bearing sand, where fierce windstorms alternately hide the grains and scatter them over the surface.

As it currently stands, the Greco-Roman descriptions of the griffin, the beautiful artifacts made in their image, and their links to the gold deposits of Central Asia could very well be field notes about a variety of dinosaurs. Until modern times, observers and writers twenty-five hundred years ago "knew" more about the real characteristics of certain quadruped dinosaurs than we did.

STEALTH BOMBS, STENCH WARFARE

The Greeks and the Romans and the societies they fought, from Hindu armies in India to blue-tattooed Brits, might have lacked twenty-first-century technology. Nevertheless, they deployed a stomach-turning spectrum of dirty bombs, chemical tricks, stealth missiles, and weapons of substantial destruction that would impress today's military commanders and terrorists no end.

Take the Scythians, those horse-happy nomads on the steppes of what is now the Ukraine and the north shores of the Black Sea. Superb archers, they possessed reflex bows (given extra power with composite materials) that allowed them to shoot nearly twice as far as their Greek counterparts. Their accuracy was legendary—and legend has been confirmed by archaeological finds of skulls with Scythian arrows embedded right between the eyes.

That edge was just for starters. Scythian warriors also sported decorated holsters full of barbed arrows dipped in the most terrifying poisons. (Some researchers think they carried small containers for their designer toxins, allowing them to anoint the arrow tips just before firing and avoid the disaster of self-envenomation.)

Most Scythian poison recipes were based on snake venom, amped up by the addition of feces, human blood, and rotting snake meat. A Greek or Roman soldier hit by a bacteria-laden Scythian arrow would quickly go into shock, then die in slow agony from gangrene and tetanus. To add to the torture, these vile concoctions carried a powerful stench, generating fear (and probably upping the desertion rate) among the troops being fired on.

The psychological power of stench warfare was itself another tactic. Although the ghastly reek of poison arrows could be counted on to terrify, long-ago fighters also flung dung and decaying corpses on enemies, and defending cities sometimes poured boiling urine on their attackers. Traditionally, the two vilest stenches in the world were those of excrement and rotting bodies—where the worst pathogens were (and still are) found.

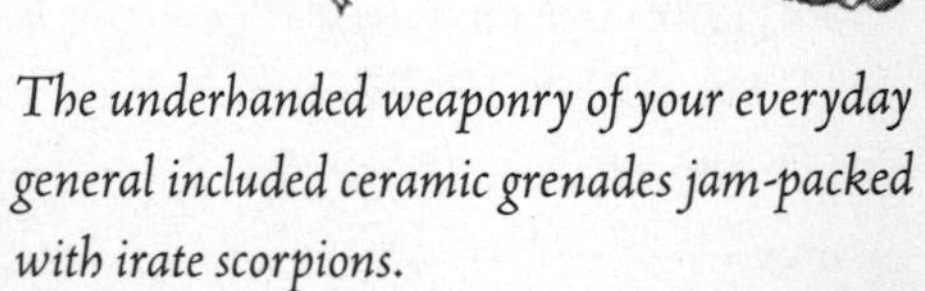

The underhanded weaponry of your everyday general included ceramic grenades jam-packed with irate scorpions.

The ingenuity of arrow-delivered toxins extended to plant material. Deadly wolfbane, hellebore, yew, hemlock, and strychnine were the top five plant-based additives, although long-ago herbalists routinely harvested dozens of plants with the power to kill. These field pharmacologists understood how best to extract and prepare the parts of the plants in which they resided.

Fire also offered scope for chemically enhanced weaponry. Arrows could be fire-tipped. So could javelins, whose heads were wrapped with pitch or other combustible materials, ignited, then launched with a spear-throwing machine or a siege engine. Originally a weapon developed by the Iberians of Spain, the *falarica* was adopted by the Roman military. Each *falarica* was a javelin that carried an armor-piercing tip. When the large fiery missile struck a soldier's shield or armor and set it ablaze, he was forced to strip away his protective gear. With its size and speed, the *falarica* could take off a soldier's arm or leg.

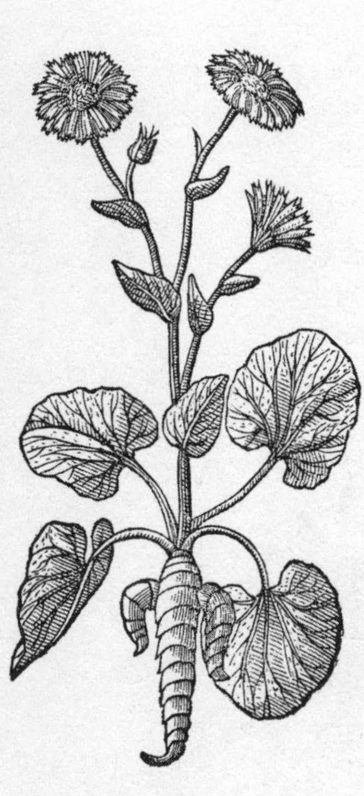

Along with snake venom and rotting flesh, combatants regularly used plant toxins like wolfbane to amp up the nastiness of their arrows.

Perhaps the most terrifying military firepower was developed by the Phoenicians. In 332 B.C., when Alexander the Great and his army began their siege of the fortified island city of Tyre, his opponents converted one of their largest two-masted vessels into a floating incendiary bomb by loading it with flammable material, including four cauldrons filled with petroleum and other substances. The Phoenicians crashed their ship onto Alexander's siege machines and the pier he'd built to get from the shore to the city, causing many casualties.

Despite the havoc this caused, Alex's forces didn't give up. The frantic Phoenicians improvised another fiendish weapon, one that uncannily foreshadowed shrapnel and other modern metallic incendiaries. Using huge pans of bronze set over charcoal fires, they heated quantities of fine sand and metal fragments until red-hot, then catapulted it over their walls. It rained a peculiarly horrible sort of death on the Macedonian soldiers, the hot grains sliding beneath their armor and causing many to go mad before expiring.

From Aeneas to Herodian, Greek and Roman historical accounts are crammed with descriptions of military weapons and defensive tactics, from the use of vinegar as a fire retardant to scorpion hand grenades—the latter launched by the Hatrans to defeat the Roman emperor Septimius Severus in the Mesopotamian (now Iraqi) desert.

But chemical warfare began way back in the fifth century B.C., between the Spartans and the people of Plataea, a Greek city-state near its ally Athens. As described by the historian Thucydides, the Spartan army had spent fruitless months besieging the city, which resisted with ingenious tactics, including the remarkable feat of lassoing enemy siege engines, winching them into the air with pulleys, then letting them crash into rubble.

At this juncture, the Spartans spent months heaping a great quantity of timber against the walls of Plataea, wetting it down with a generous amount of pine sap. Before setting it afire, they added sulfur. The hungry flames burned hotly, the mixture producing a poisonous gas called sulfur dioxide. Remember "fire and brimstone," that evocative phrase used by preachers of old? Brimstone was simply another word for sulfur, a useful substance but highly toxic and flammable.

At first, the defenders of Plataea ran to escape the poison gases. Before the Spartans broke down their gates, however, the winds shifted and a thunderstorm hit, quenching the flames. Even the Spartans knew better than to question the intervention of the gods, which to their minds this clearly was.

Centuries later, in a set-to between Romans and Persians, a similar scenario had a different outcome. The Romans had conquered the city of Dura-Europa on the Euphrates, installing a military garrison there. In A.D. 256, the Persians launched a siege to regain the city. Both sides dug tunnels, trying to undermine—or protect—the city walls. At one point, the Persians fired up a brazier, stoking it with bitumen and sulfur crystals, then used bellows to send deadly gases toward Roman soldiers through the underground chamber. Recent archaeological excavations found the bodies of twenty men who'd died of asphyxiation, not by sword or spear, showing how effective the toxic gas attack was.

In ancient times, negotiating adversaries would often agree on principle not to use chemical weaponry, biological warfare, and other ignoble tactics. As the foregoing proves, more often than not, short-term goals and the convenient amnesia of human memory won out.

THE CITIZEN WHO BARKED AT THE WORLD

Diogenes of Sinope, a Greek city-state on the south coast of the Black Sea, came by his philosophical calling in an offbeat way. As a young man, he and his banker father got embroiled in an embezzlement scandal. Details are contradictory, but Diogenes, who may have been adulterating the local coinage, was promptly exiled from his native shores.

He and his one-way ticket ended up in Athens, where he drifted into study with Antisthenes. Diogenes' teacher was a Cynic, a new breed of radical thinkers whose snarling behavior, satirical spirit, and blunt questions attacked convention and the status quo. Diogenes fit right in. After arguing with his new mentor, who smacked him, he became a stray dog of a street person. Eating whatever he begged or found, he relieved his bowels in public places and got a reputation for urinating on those who criticized him.

Outside the Metroon, a civic building in the agora or marketplace, Diogenes stumbled across an old pithos, one of the oversized earthenware storage jars the Greeks used to hold grain or wine. Homeless no more! He immediately moved in.

A cultural aside: accounts of Diogenes often refer to his dwelling as a tub or a jug. Granted, it wasn't a double-wide. Nevertheless, his pithos provided a snug home for a man—even a house guest on occasion. Despite his refusal to partake of money or hygiene, Diogenes attracted admirers. And overnighters. Lais, a high-priced hetera and artists' model, chose him for her pro bono work. A gal who enjoyed the give-and-take of big questions about the universe, she worked out a time-share arrangement with Diogenes and a well-to-do philosopher named Aristippus, who was forced to shell out her going rate.

Since Diogenes had already been accused of tampering with the money minted in his hometown, Sinope, he decided to carry the metaphor further. He began to lecture the Athenians about their pious hypocrisy, when real evil still existed in the world. "Alter your currency!" he would thunder. A saucy wit, Diogenes had a comeback for almost any remark. With his austere lifestyle, he could laugh at pretensions. After he saw a peasant boy drink from his own cupped hands, Diogenes got rid of his sole utensil—a wooden bowl.

Cynical about people but never about dogs, Diogenes fed local canines his octopus leftovers.

The Cynic outraged almost everyone—and especially with this twenty-first-century-sounding remark. Asked where he was from, Diogenes replied, "I am a citizen of the world." This brand-new coinage, supposedly the first use of the word *cosmopolitan*, was shocking to Greek ears. In those days, each person identified strongly with his or her city-state. To reject your own city-state (even if it had rejected you) and claim the world was obscene.

Diogenes allegedly wrote diatribes and books called *Jackdaw*, *On Love*, *The Mendicant*, and others, but none survive. Besides shaking up complacent ideas of conventionality and self-sufficiency, he gave writers and historians bagfuls of juicy stories to chuckle or tsk-tsk over.

One well-known anecdote: When Alexander the Great made an appearance in Athens, he wanted to meet the sharp-tongued philosopher whose antics had everyone buzzing. Heading toward the pithos near the Metroon, he found the philosopher seated on the ground. Introducing himself, Alex, in Great-Man mode, said, "Go ahead—ask me any favor you like."

Diogenes looked up from his sunbathing. "Stand out of my light."

Later, eager to show what a cool guy he was, Alex would tell everyone, "Had I not been Alexander the Great, I would've liked to be Diogenes."

Years after this, Diogenes went on a voyage to the nearby island of Aegina. On the way, he and others on the ship were captured by pirates. The buccaneers took him to the slave-market port of call on the big island of Crete and wholesaled him. A low moment, one would think. As usual, the irrepressible side of Diogenes came to the fore. When asked by the slaver in charge of retail sales what skills he had, Diogenes replied: "I know how to govern men." Then he added, "Ask if anyone would like to purchase a master for themselves—I'm available."

Xeniades, a wealthy Corinthian with a sense of humor, happened to be browsing in the slave market. He bought Diogenes to teach his two sons, and the men sailed back across the sea to Corinth. Before long, Diogenes' quirky ways and evident brilliance attracted other young pupils to the Xeniades household. For some time, he instructed the boys in history, poetry, and philosophy, as well as taking them hunting.

In the year 323 B.C., he died in Corinth. (One ancient biographer claimed that he died on the same June day as Alexander the Great.) On his grave, the citizens of Corinth set up a fancy pillar; at the top they placed a dog carved from Carian marble.

As with other celebrities of ancient times, tales quickly sprang up as to the manner of his death. One report had it that the ninety-something philosopher held his breath until he expired. Another said he got colic from eating raw octopus. The version that sounds most like Diogenes declared that while dividing a raw octopus among the stray dogs he regularly fed, Diogenes got badly bitten and died of infection.

ROGUE REINCARNATOR

Oh, to be famous. Better yet, infamous. That gusto for undeserved glory, that fever for instant fame, that sense of entitlement for one's own time in the spotlight. A twenty-first-century affliction, yes, but also an obsession more than twenty centuries ago.

One striver for deathless recognition was a man who called himself Proteus Peregrinus—which roughly translates as "number one wanderer." Born about A.D. 95 in the Greek town of Parium in Asia Minor, he majored in minor scandals while waiting to come into his inheritance. Losing patience with his dad's longevity, the teen took action by strangling his father. Thereupon, the young suspect fled to Palestine and joined a Christian group, considered at that time by Roman authorities to be an up-and-coming nuisance cult.

Proteus had the qualifications to be an outstanding charlatan, soon gaining a role as their main spokesman. When Proteus got tossed in jail, his Christian followers, thrilled with his martyr potential, sprang into action: breakout attempts, banquets in his cell, and, best of all, a stream of donations.

Palestine's discerning governor, however, saw the kid as a manipulator rather than a martyr. He released Proteus. Deeply chagrined, the young man abandoned his adoring flock and hitchhiked to his hometown, where the heat was still on regarding Dad's still-unsolved death. Instead of picking up his inheritance, the quick-witted lad became a Cynic philosopher on the spot. Already grungy and long-haired, it was a simple matter to tell the assembled citizens of Parium that he had adopted the Cynics' austere lifestyle and was giving them all his worldly goods, meaning Dad's farms and money.

Amid cheers, Proteus walked away from any future problems with parricide accusations, and still had some running money left from his Christian fans.

Off to Egypt, where he studied for ages with an ascetic wacko named Agathobulus. Eventually, though, the disciplines of shaving half his head and encouraging others to whip his hiney with fennel stalks proved too arduous. Proteus wandered north to Rome, trying to win fame or at least a market share of negative attention by verbally abusing the current emperor.

Expelled as just another two-bit troublemaker, Proteus wearily made his way to Greece, where he finally managed to connect with people by enraging them. First he slandered the folks of Elis, home city of the Olympic Games; then he managed to rouse a number of Greeks to take up arms against the Roman oppressors. Now he was on a roll rivaling anything that Diogenes, that disgraceful old Cynic, had done.

But he wasn't getting any younger; already in his fifies. Proteus pulled himself together to attend the weeklong Olympic Games, determined to commit an outrage that would make his name forever. Once there, he noticed the new water and sanitation system, courtesy of an altruistic benefactor named Herodus Atticus. Perfect target. He really got a rise out of the mostly male crowd when he roared that they'd become effeminate. "What next? Public toilets at the Olympic Games? You oughta be ashamed!" All this orating made Proteus thirsty, so he drank from the newly installed fountain before continuing.

At the sight of Proteus enjoying the water he'd just maligned, the crowd started pelting him with stones. He ran to the temple of Zeus for sanctuary, tickled with his reception and already plotting his next career move.

At the next Olympics, he reversed course and praised Atticus, the man he'd libeled for so long. The crowds, anticipating a vicious attack on something or someone new, melted away. "The guy has gone stale," everyone said.

But they hadn't heard the last from Proteus Peregrinus. After intense thought, he came up with a plan to ensure his undying fame. Only one minor hitch: it required him to commit suicide.

At the next Olympic Games, Proteus (now approaching geezerhood) publicly announced his decision to kill himself in a way that displayed really awesome fortitude. His plan was to immolate himself at the height of the festival in a spot where the maximum number of attendees could see and smell him burn. A crowd of fifty thousand people would make a very decent group for his exit poll.

Wouldn't you know it—scandal-averse games officials intervened and forbade a Proteus roast on their hallowed grounds. He was forced to make other arrangements off-site.

Late one moonlit night after the finale of the Olympic Games, on a hill two miles from the sacred precinct, the number one wanderer took his last stroll.

Lacking the funds for a more lavish cremation, he'd ordered the standard funeral pyre, a pit six feet deep filled with torchwood and brush.

Dressed in a filthy shirt and looking pale, Peregrinus gave his own funeral oration and announced that he had, in the finest tradition of celebrities, changed his name to Phoenix. He threw a handful of incense on the blaze, hesitating coyly.

A few in the crowd called out, "Preserve your life for the Greeks!" They were, however, shouted down by others, who chanted, "Carry out your purpose!"

Mr. Call Me Phoenix threw himself into the blazing pyre and was enveloped.

The author Lucian of Samostrata, eyewitness to this weenie roast, later wrote, "So ended that poor wretch Proteus, a man who never fixed his gaze on the verities, but always did and said everything with a view to glory and the praise of the multitude, even to the extent of leaping into fire, where he was sure not to enjoy the praise because he could not hear it."

Proteus meant for the manner of his death to echo the reincarnation faith of the philosophers of India, but he lacked their integrity. Several had immolated themselves before Greek and Roman audiences—the most famous being the gymnosophist Calanus, who calmly climbed onto his own crackling funeral pyre with Alexander the Great and his men as witnesses. The suicidal glory hound once briefly known as Phoenix—and now not known at all—failed to rise from the ashes. Surely there's a lesson here for more recent wingnuts.

SET FREE BY STOICISM

Born into slavery or sold into it around A.D. 65, Epictetus came from a small town called Hierapolis (Pamukkale in present-day Turkey). As a slave and thus an object to be bought and sold, he ended up in Rome. His name simply meant "acquired."

His owner was a freedman—normally a circumstance that would create terror in a slave's heart, since all too often former slaves made the harshest masters. This man, however, happened to be Epaphroditus, a freedman who'd amassed wealth and power working as a government official for none other than Emperor Nero.

A Neronian association would usually inspire even more abject terror in a slave's heart. Chalk it up to altruism or simple goodwill, but Epaphroditus saw something special in his slave—charisma, intellectual curiosity, potential. Moreover, the boy was crippled in some fashion, possibly from childhood or from the rigors of enslavement.

The master paid for his slave to study with Musonius Rufus, a well-known Stoic philosopher of the first century. The young slave took to it like feta to Greek salad. He readily absorbed the teachings of the early Greek Stoics, including Zeno and Chrysippus. Having gained instruction in moral philosophy and literacy, Epictetus began to write his own handbook on living the virtuous philosophic life. In ancient times the word *virtue*, often thrown around in such discussions, meant something closer to "excellence" in Greek.

At some point, Epaphroditus freed Epictetus. The freedman may have wished he'd pursued philosophy himself when, in A.D. 68, Nero's number came up. Deserted by his staff, friends, and family (not that many were left, after Nero's bloody purges), the emperor fled in terror when the Roman Senate put him on its most-wanted list. Loyal to the last, Epaphroditus accompanied Nero in flight—then had the nauseating task of helping the emperor slit his throat when he couldn't do it himself.

Then as now, bureaucracy moved slowly; Epaphroditus lived a normal life

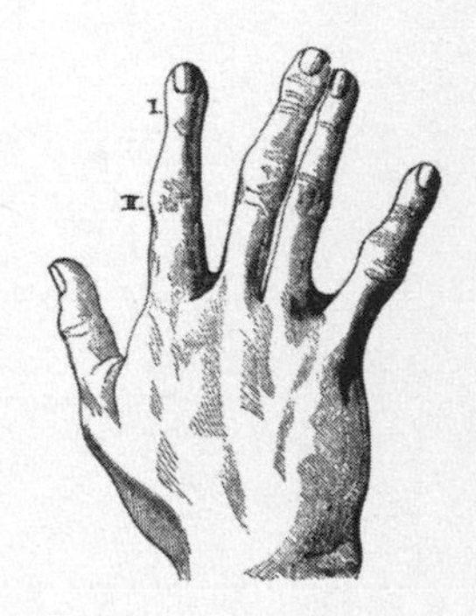

Centuries apart yet on the same philosophical path, these two men handed out written advice that still resonates.

for a dozen years until someone remembered his association with Nero and he was put to death by Emperor Domitian in A.D. 81.

In the grip of a growing paranoia, Domitian eventually exiled or executed a variety of senators and military men, then banished all philosophers from Rome, including Epictetus, who left to open a school of his own in northwest Greece at Epirus. Although the huge commute left something to be desired, he attracted many Roman patricians to his academy. Around A.D. 104 he, or perhaps Arrian, one of his enthusiastic followers, wrote a book called *Discourses* (plus a small CliffsNotes abstract of the book), which has survived to modern times.

Candid to the last, Epictectus left behind commonsense instructions for living a good life that still resonate today. "Seek not what you want to happen," he said. "Seek to want what happens."

Fast-forward to A.D. 161, when the deathless words of Epictetus gave great consolation and moral courage to Marcus Aurelius, called "the last of the good emperors." Poor Aurelius, one of the best and brightest, was forced to spend fifteen of his twenty years as emperor fighting ghastly wars under wretched conditions with the Parthians, the German tribes, and worst of all the plague, which the Parthians bestowed on the Roman army as a parting shot.

As he alternately froze and roasted on the northern frontier, the fiftysomething emperor took a daily dose of theriac, a nostrum laden with opium, to ease the pain of his stomach troubles—possibly cancer. To keep sane, Marcus Aurelius wrote in his notebook, which he'd labeled "To Myself."

An excerpt: "The first rule is, to keep an untroubled spirit; for all things must bow to Nature's law, and soon enough you must vanish into nothingness, like [Emperors] Hadrian and Augustus. The second is to look things in the face and know them for what they are, remembering that it is your duty to be a good man. Do without flinching what man's nature demands—say what seems to you most just—though with courtesy, modesty, and sincerity."

Marcus had absorbed well the teachings of Epictetus and the lessons of Stoicism, while adding eclectic insights of his own. Today's seekers of philosophical happiness are doubly lucky to have his notes—now a book called *Meditations*—and that of Epictetus to treasure.

The pitiful irony of Marcus Aurelius' life was the aftermath. Although he and his wife, Faustina, had fourteen children, only one son lived through childhood. Wouldn't you know it, that son turned out to be Commodus—one of Rome's most deranged and vicious rulers ever. It was enough to make the staunchest of Stoics weep into his resinated wine.

The wise and philosophical words of an emperor and a slave, connected through time, have been the inspiration for myriad books and films, including Tom Wolfe's *A Man in Full* and the historically inaccurate but moving film *Gladiator*. Perhaps the most poignant use of this brand of Stoicism occurred during the Vietnam War, when fighter pilot James Stockdale leaned on the words of Epictetus to help him endure seven and a half years of imprisonment, torture, and solitary confinement.

INCENSE OPEC

Gold invariably got people's attention, but what the ancient world craved—and consumed in mind-boggling quantities—was the begrudging output of two scraggly, thorny trees.

Meet frankincense and myrrh, two of the three treasures given, as you might recall, to a certain babe in a manger. The Greeks believed that their deities had superpowers, including sensory ones. To please the nostrils of the gods and goddesses, they burned frankincense (literally, "true or choice incense") upon their outdoor altars. Its fragrant white smoke was also used to please the nostrils of ordinary mortals, to fumigate homes, clothes, latrines—and to mask the ghastly stench of sickrooms reeking of pus and gangrene. Frankincense healed sores and eased aching teeth and inflamed eyes. Its bark was touted as a sovereign remedy for people who coughed up blood when drunk. Troubled by warts? Powder of frankincense, mixed with vinegar and pitch, would do the job.

Myrrh, on the other hand, was too valuable to use as incense. Its tear-shaped reddish-brown beads were employed to fumigate, to embalm, and as an additive to anointing oils. Long-ago doctors employed it for a variety of ills, including mouth sores, ulcers, and bronchial complaints. Tiny beads of it were burned as flea repellent. A major ingredient in many medicines, myrrh's virtues as a painkiller have been proven in our time—two of its compounds have strong analgesic effects.

This puny plant is the famed myrrh whose bitter perfume relieved pain and kept wine from souring.

Another key arena: winemaking. The science of keeping wine from turning got a real boost when Greeks discovered that myrrh halted the bacteria that produced vinegar while doing no harm to the yeastie beasties that produced alcohol. Other aromatics, including cinnamon, lavender, cedar, and peppercorns, were used as additives but myrrh worked best. Myrrh wine also served as a wound dressing, allowing patients to enjoy their wine topically as well as gastronomically.

There were downsides. Neither plant could be domesticated, and they grew wild in just two remote places—the Horn of Africa region of Somalia and the southern Arabian coast. Paralleling the petroleum industry of more recent times, the production of frankincense and myrrh was strictly controlled by an OPEC of old: the five incense kingdoms, called by encyclopedist Pliny "the richest people on earth."

At first the Hadhramawt rulers in Arabia had a lock on incense; by the second century B.C., however, the rival kingdom of Saba began to exploit Somali frankincense and ship it by sea. This did little to ease the economic stranglehold.

It took a complex network of gatherers, merchants, camel drivers, and myrrh middlemen to collect and transport the resins, which were taxed at sixty-five separate camel stops along the Incense Route. To add to the difficulty, the historian Herodotus noted, the pesky plants were guarded by winged serpents, which could be driven away only by the smoke of storax, another fragrant resin. Herodotus, who loved a good story, clearly couldn't resist including this one in his book—a fable that enabled sellers to charge even more for their crop.

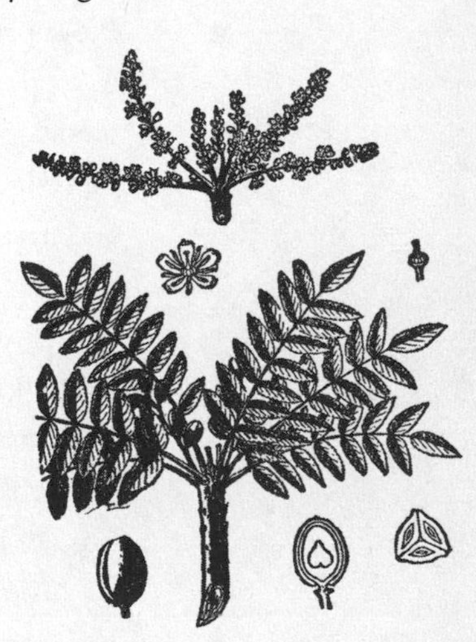

The golden beads of resin harvested from the scraggly frankincense were lavished on altars and funeral pyres alike.

Snakes aside, collecting the resins was arduous because the plants grew in desolate, fiery-hot wilderness. Collectors were ill-paid natives (as they continue to be today), subject to the intense scrutiny nowadays shown to diamond workers. To tap frankincense, they made incisions in the bark, then harvested the gum fifteen days later. The highest-grade myrrh was that which oozed naturally from the trunk; trees were later tapped for a second harvest.

In the year A.D. 65, Emperor Nero and his profligate ways nearly threw the sweet-smelling incense economy into a tailspin. That summer, aggrieved over having kicked his pregnant wife, Poppaea, to death in a fit of road rage ("I told her not to nag me about coming home late from the chariot races, but she kept on—

she made me do it!"), Nero planned a funeral that would outdo those of Alexander the Great and Julius Caesar combined.

For her send-off, he ordered a whopping amount of incense. Just how big would whopping be in this case? The amount shipped to Nero was that year's entire harvest from Arabia. And every single fragrant golden bead was ladled onto Poppy's funeral pyre in Rome. Call it an asthmatic's nightmare: not much genuine grief but a whole lotta smokin' going on.

HARD-TO-HANDLE HYDROCARBONS

Now it can be told: by A.D. 50, folks had discovered fossil fuels *and* the secret to tapping into them by exploiting a heretofore useless product: menstrual blood. Jewish-Roman writer Josephus described the painstaking process carried out at the Dead Sea, then known as Asphalt Lake: "The waters are bitter and unproductive . . . In many parts the lake casts up black masses of bitumen, which float on the surface, in their shape and size resembling decapitated bulls. The laborers on the lake row up to these and, catching hold of the lumps, haul them into their boats. But when they have filled the boats it is no easy task to detach their cargo, which owing to its tenacious and glutinous character clings to the boat until it is loosened by the monthly secretions of women, to which it alone yields."

In order to harness this ominous periodical power, local collectors would've had to make prior arrangements with a number of obliging women. Think of it. Had fossil fuel addiction caught on in a big way, women could have won financial independence millennia ago by marketing—or withholding—something that showed up as regularly as moonlight among females of breeding age.

Tragically, no one at the time had much enthusiasm for hydrocarbons, whether you called them bitumen, asphalt, naphtha, or inflammable mud. Too volatile, too smelly, too hard to ship or store. Neither chunks of bitumen nor thick, sulfurous fossil fuels worked very well for the everyday needs of heating and lighting in the furnaces, braziers, cookstoves, and lamps of Romans, Greeks, Jews, and other contemporaneous cultures. Most people stuck with standard olive oil.

One bit of evidence to the contrary remains in the obscure writings of a Byzantine reporter named Condinus. According to him, during the tenure of the Roman emperor Septimius Severus (A.D. 193–211), two large thermae (public baths) were built, one of them accommodating two thousand bathers at a time. Although for centuries Roman baths had been heated with wood fuel via an underfloor hypocaust system, Severus boldly chose to break with tradition in this instance. Inside the structure, both the air and the bath waters were heated by a mysterious petroleum product called "Median fire." The heat source was supposedly contained in glass or earthenware lamps. Archaeologists would love to solve this mystery from history, but the thermae in question evidently left no trace, as their destruction by later conflicts was so thorough.

Bitumen and asphalt weren't completely useless. These heavy tar-like forms of hydrocarbon did an excellent job of waterproofing, protecting woodwork, and caulking boats. (That's why those fellows at the Dead Sea were busy fishing out bull-shaped chunks of bitumen.) There were export sales to the Egyptians, who used tar for embalming. Got toothache or the gout? Troubled by leprous spots? A dab of hydrocarbon would do you.

Like today, fossil fuels were found in greatest abundance in the Middle East, but a few sources bubbled up elsewhere. In Sicily, for instance, a spring in the city of Agrigentum leaked a light oil that folks collected with reed brooms, then burned in their lamps. Locals reserved their highest praise, however, for the oil's ability to cure scab in domestic animals.

In ancient times, nine out of ten generals agreed that fossil fuels made exciting if unreliable weapons, given their magical quality of instant ignition. It was putting the brakes on that presented the problem. Along those lines, during a stopover in Ecbatana on Alexander the Great's world conquest tour, locals eager to please did some parlor tricks with naphtha. Turning a street into a sheet of flames quickly palled, however. Sensing the boredom of the big shots, one lad suggested, "Let's set Stephanus on fire!" The boy, a good singer but no intellectual powerhouse, agreed. After a petroleum rubdown, he burst into flame. Luckily, they managed to contain Stephanus before he burned to a crisp.

Spectator Alex had had painful prior experience with the fiery stuff himself. In 332 B.C., while besieging the city of Tyre, he and his men were attacked and burned by a suicide ship loaded with burning petroleum.

Two hundred years later, the Roman general Lucullus and his army would undergo a similarly excruciating experience. While besieging a walled city in Mesopotamia, the legionaries got doused with maltha, an inflammable mud from a local marsh. The maltha stuck to bodies and armor, literally roasting the Roman soldiers to death. Military brass would eventually find that the hydrocarbon-based mixture called "Greek fire" showed far more promise.

Although it's no longer called the Asphalt Lake, the Dead Sea is indeed closer to death in our century. Its shrinkage is mainly due to thirsty nations in the neighborhood, which use much of its freshwater inflow. As happened in the first century A.D., large chunks of black asphalt—some weighing over a ton—have bobbed to the surface of the Dead Sea, beginning in the 1950s and continuing today. Most likely liberated by seismic shifts in underwater faults, their appearance may set off a new rush for the eight to ten billion barrels of "black gold" thought to lie beneath the basin of the Dead Sea.

If that happens, perhaps those monthly secretions may come into play once again, revisiting a whole new entrepreneurial opportunity for women.

HOROSCOPES AND OTHER DISASTERS

As the years passed, astrology first edged out bird cries, lightning strikes, and entrail reading as the Roman world's most popular divination method, then soared to the top of the popularity polls. Everyone from slaves to empresses got their horoscopes read.

Although astrology got breathless attention, no one ever asked, "What's your sign?" That would have been the equivalent of a bomb threat in the airport ticket line. Your natal horoscope and date of birth were guarded like computer passwords. Identity theft? More like fate interception. With a valid DOB, any halfway decent astrologer could make a valid DOD prediction. High rollers, from emperors to newly wealthy charioteer stars, had a lot riding on such a scary eventuality.

A constellation called Canis Major ("bigger dog" in Latin) boasts Sirius, the brightest star in the sky.

Astrological readings differed from today's horoscopes by quite a margin. Long-ago astrologers concentrated on the twelve signs of the zodiac and the seven planets (five, actually, plus the sun and moon). To make matters more confusing, most of the celestial objects overhead were sloppily referred to as stars as well. The stars' lives, likes, and dislikes were followed as obsessively as the twenty-first-century fan Web sites of pop singers. Each star had its own ruling color, hour of the day, plant, animal, mineral, and list of attributes. (One remnant of that old-time astrology: the pairing of zodiac signs with gems.)

Astrology had been around for centuries in the Middle East, especially in the southern part of Babylonia called the Chaldea district—to such an extent that *Chaldean* became a Greco-Roman synonym for *astrologer*.

Greek astrologers added a twist to the Babylonian variety by establishing the importance of the zodiac sign rising on the eastern horizon at the moment of a person's birth. The word *horoscope,* meaning "hour watcher," refers to that key moment.

Romans didn't take to astrology straightaway; rather, the public took to it but leaders feared it. Consequently, there were periodic expulsions of astrologers from the city, the first one in 139 B.C. Like the periodic sweeps of prostitutes made by law enforcement, it wasn't long before things quietly returned to the prior relationships between clients and astrologers.

The mainstream popularity of astrology got a tremendous kick-start from a polymath Syrian Greek named Posidonius, who traveled widely and studied earnestly, mastering astronomy, mathematics, and other disciplines. Above all, he was addicted to astrology, writing five books on it. About 90 B.C., he meshed all his insights and learning into a pleasing worldview called the doctrine of the unity of the cosmos. To him, mankind reflected the universe. Free will and chance? Out the window; determinism and acceptance of fate were in, interlocking well with the stiff-upper-lip attitudes of Stoic philosophy. His brand of astrology sounded awfully good to the ruling classes, since it meant that the stars dictated fate, keeping slaves, women, and other troublesome elements in their predestined places.

Although Posidonius had a couple of minor scientific epiphanies to his credit, his lasting influence went light-years beyond his actual accomplishments. The man had fatefully great timing. A splendid speaker, interesting and well-traveled, he rubbed elbows with the celebrated, won the friendship of the high and mighty from Cicero to Pompey the Great, and got praised by later writers and astrologers.

Some Roman leaders were spooked by astrology's death predictions and took steps to counteract them. Other rising stars took the opposite tack. When Julius Caesar came into power, for instance, he scoffed at predictions but did nothing to discourage Nigidius Figulus, Rome's leading astrologer, who gave JC a prophecy of a lucky new era. In the power struggles and civil wars that raged after Caesar's death, Marc Antony made use of astrology for his own ends.

But the most ingenious use of astrology was devised by Marc's ultimate opponent, Octavian. Once the smoke had cleared and Octavian finally became Rome's first emperor at age thirty-six, he had coins struck with his birth sign on them—with one sly amendment. The coins said Capricorn, but his real sign was otherwise, his birth date being September 23.

Despite brief intervals of official disapproval and expulsion, astrologers continued to find gainful employment at all levels of Greco-Roman society for centuries to come.

An ancient astrological blueprint has survived to our times, a veritable idiot's guide to the working techniques of long-ago forecasters—and a grand source of amusement as well. Written by Vettius Valens, another Syrian astrologer and a younger contemporary of megawatt stargazer Claudius Ptolemy, it was a comprehensive textbook with generalities about astrological forecasting, followed by 125 examples of actual horoscopes cast. Valens also went into detail as to how each star rules for a given period, then cedes control to another, in the process showing just how

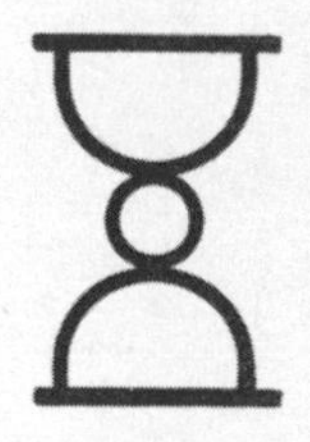

Skygazers wrote horoscopes based on the hour of birth, using this hourglass symbol.

the prognosticators of old bought themselves ample wiggle room for their predictions.

Here's a much-abridged excerpt from Valens, showing the influence of the heavenly body of Saturn: "Saturn makes those born under him petty, malignant, care-worn, self-deprecating, solitary, deceitful, secretive in trickery, strict, squalid, black-clad, sad-looking, with a nautical bent, plying waterside trades. Saturn also causes humblings, sluggishness, unemployment, obstacles in business, interminable lawsuits, secrets, accusations, tears, capture, exposures of children. Saturn makes farmers because of its rule over the land, and causes men to be renters of property, tax farmers, and violent in action. Of materials, it rules lead, wood, and stone. Of body parts, it rules the legs, knees, lymph, phlegm, bladder, kidneys. Of syndromes, it rules possession, homosexuality, and depravity. Saturn makes bachelors and widows, bereavements, and childlessness. It causes violent deaths by water, strangulation, imprisonment, or dysentery. It also causes falling on the face. It is the star of Nemesis, of the day sect. Like castor in color, it is astringent in taste."

All the elements for a good soap opera, in one astringent astrological sign. Puts our syrupy, soothing syndicated newspaper horoscopes in the shade, doesn't it?

THE ASTROLOGER-EMPEROR CONNECTION

Tiberius Claudius Balbillus had to admit it felt grand to come from such a celebrity-rich line of celestial investigators. His own dad, Thrasyllus, had been personal astrologer to Emperor Tiberius. He had ties to minor nobility, even if his great-great-granddaddy had ruled over a corner of Syria that only a mountain goat could love.

He'd been lucky, befriending that goofy, drooling kid at Emperor Octavian Augustus' palace back when they'd both been young. Who'd have dreamed that Claudius would become emperor at age fifty? So much for royal horoscopes—and he, Balbillus the Wise, was supposed to be a professional at them.

His early childhood in Alexandria had been golden, studying philosophy and the classics. Then an equally enjoyable stay in Rome, until that young wingnut Caligula got all paranoid about astrologers. Four years later, the goddess Fortuna smiled again—suddenly young Caligula was lying on a funeral bier, and Claudius was sitting on the throne.

Right away, the new emperor Claudius thought of him. A huge honor, getting to accompany old Claw-claw as an officer in the Twentieth Legion; a pity that their junket had to be to that Druid-infested island called Britain.

Still, upon their safe return to Rome he got a special crown from Claudius—and an enviable appointment as director of the Great Library and Museum of Alexandria. Back to Egypt! Not too shabby, and a priesthood besides.

He loved how the job required him to take periodic trips between Alexandria and Rome. Deluxe transit, of course. One of those superships. And then, striking pay dirt with that eclipse forecast he made! Too bad the darned thing happened to coincide with August 1, the birthday of Emperor Claudius. He'd had to give his boyhood friend some intense counseling. Never did yank the emperor out of his depression, but at least Claudius didn't expel him, like he did every other astrologer in Rome.

This symbol meant "day."

Such a shame when Claudius died. A shock to almost everyone except his wife, Agrippina, and that monstrous pimply-faced blond kid of hers.

He'd had his hands full, ascertaining how to suck up to the incoming administration.

He and young Nero got along fine. More than fine. At the first crisis, when the kid got terrified about a comet, he, Balbillus the Wise, gave him the solution. A scapegoat or two. Cut their horoscopes off at the knees. Might as well do some judicious pruning in the Roman Senate while he was at it. Nero went right along. Such fervor—a teen who really took to mayhem. Turned into a bit of a tightrope at the old palace, what with the emperor plotting to kill off the rest of his advisors and family.

Luck held again, though—he cajoled Emperor Nero into giving him another appointment back in sunny Egypt, this time as prefect. Great job—plenty of time to cast a few horoscopes, catch up on his reading, muck about in the sand, maybe dig out around the funny old Sphinx a little more.

How time flew. It was A.D. 69 already, Nero and his three successors were dead, and suddenly his other old pal Vespasian was Roman emperor! Once again, Balbillus was summoned for honors and a place by the side of the new first man in Rome.

His star signs had always been fortunate. Even though he'd only had the one offspring, and she a girl, she'd married well—another Greek prince from the Commagne of his forefathers. It was high time he sent his spirit up to live with the stars—see if all the astrology nonsense he'd practiced his entire life was true.

This one signified "day and night."

The luck of Balbillus continued to hold after death. His splendid brain and even more dazzling kiss-up abilities would pass unimpaired to his heirs. His daughter, Claudia Capitolina, would produce a grandson and a granddaughter who would both grow up to achieve long and marvelously sycophantic relationships with Emperor Hadrian and Empress Sabina.

And his friend, Emperor Vespasian? To honor Tiberius Claudius Balbillus, in A.D. 79 he would establish an athletic festival in Ephesus called the Balbillean Games. The games to honor this charming rogue of an astrologer continued to be held in that city for two hundred years or more.

This symbol expressed the word "year."

SECTION IV

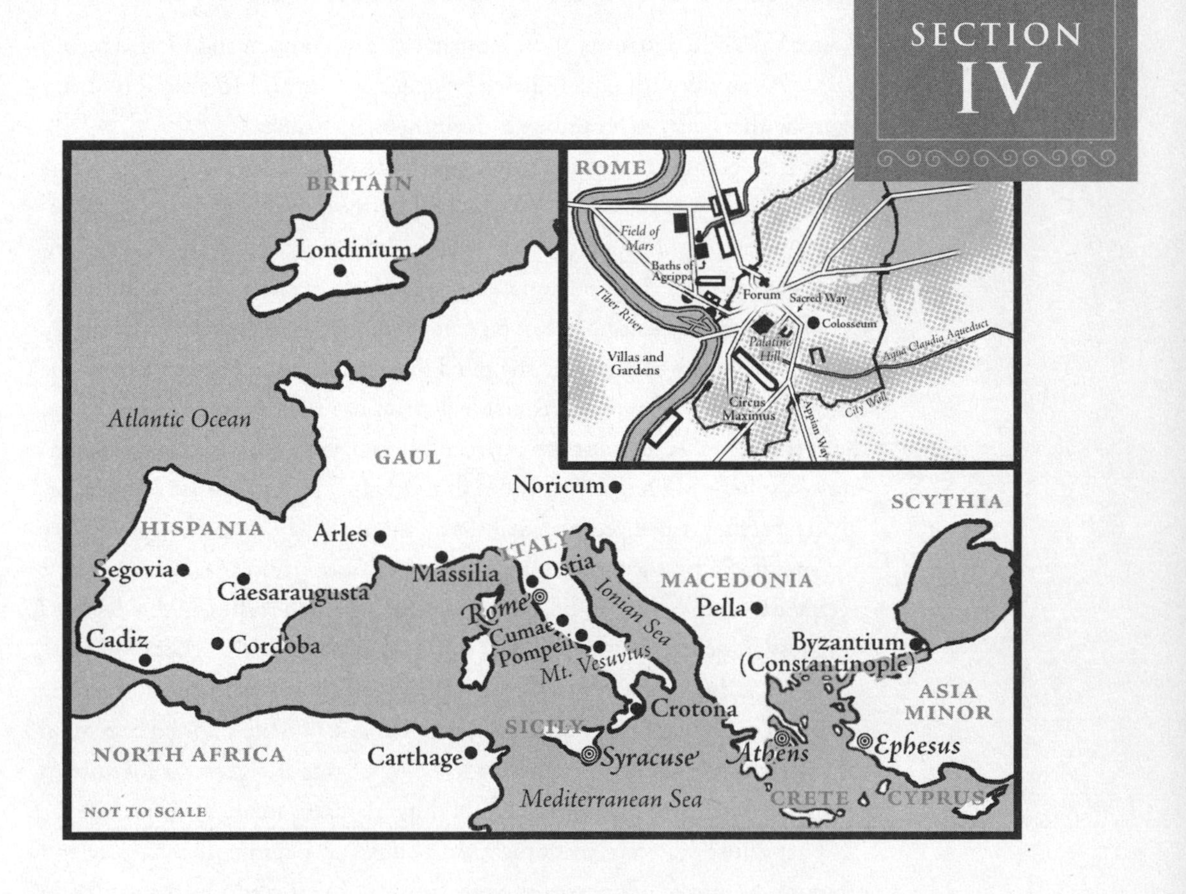

Rome & its Provinces

SECRETS OF ROME'S HEAVENLY MUD

Mount Vesuvius, destroyer of the Roman cities of Pompeii and Herculaneum in A.D. 79, could with equal justice be called a creator. How so? The story begins with a concrete example: the Pantheon of Rome.

At first glance, this venerable landmark sometimes disappoints. The Pantheon hunkers down in a piazza, half suffocated by neighboring structures. As you pass through its bronze doors, however, prepare for visual glory. Its implausible interior soars up to a single cerulean blue eye in the heavens, its hollow dome appearing to rest on pairs of tall, time-mellowed red columns. Gazing at the alcoves that once sheltered the seven heavenly gods, you're at the center of a knowable universe, inside a structure that reduces humans and exalts them, too. A temple both open and closed, its height matches its diameter—a sphere 142 feet across could fit neatly inside. It's the first monumental architecture conceived primarily as an interior.

After its first incarnation, built by Marcus Agrippa in 27 B.C. as a modest rectangular temple, the Pantheon has twice risen from the ashes. In this, its third and most glorious do-over, the Pantheon has logged more than 1,880 years.

To what does it owe its longevity? Luck, certainly. And more luck in the Christian era, being repurposed as a church in A.D. 608 but allowed to remain in its Emperor Hadrian–inspired form. What was the greatest fortune to befall the Pantheon? The choice of building materials made by its architects.

They chose concrete, most plebian of construction ingredients, gray and grainy and unloved. Concrete, the one substance that could be persuaded to form a perfect hemisphere, a solid dome weighing 5,000 tons, 20 feet thick at the widest diameter of the circle, narrowing to 7.5 feet at the top, all of it held up, seemingly, by nothing.

This was no ordinary concrete, the kind from which shoddy twentieth-century buildings and quick-to-fracture sidewalks are made, nor was it braced with invisible iron rebar. Its architects built the Pantheon with celestial concrete, fit for the Roman gods, made with one part hydrate lime to two parts pozzolanic ash, a fine-textured sandy residue from the mountainous area that included Mount Vesuvius.

The dome of the Pantheon, temple to all gods, was deliberately built with its great blue eye open to the heavens.

It's taken engineers, architects, and concrete producers millennia to unlock the mysteries of Roman superconcrete. The first secret: minimal amounts of water to make it. This worked for the Romans because they mixed each batch in a mortar box with a special hoe, the same way they made mortar for bricks. No fluid pouring; this was thick, "no slump" concrete.

Second, as the Romans learned via years of experimentation, the ingredients set off a chemical reaction that worked down to the atomic structure level. Unseen by human eyes, the calcium hydroxide of the wet lime slid readily into the molecular holes of the silica in the pozzolanic ash. The result? An amazing gel that permanently bonded all components together. As an encore, this concrete also performed its sleight of hand underwater, as witness the maritime harbor remains at Cosa, Italy, securely in place since this hydraulic mortar was invented.

The third secret? A killer application. Workers took the thick, stiff concrete mud, slathered it over a layer of aggregate rock pieces, and then proceeded to pound the concrete mortar into the rock layer. Researchers in the twenty-first century have discovered that tightly packing the concrete mortar not only drives out excess water but actually produces more bonding gel. The famed writer-architect Vitruvius knew about pozzolanic ash and the importance of tightly packing concrete in the first century B.C., mentioning the technique in his book *On Architecture*. By the time the Pantheon was built, the technique was standard operating procedure. For the Pantheon, its builders used heavier aggregate, such as travertine, for the walls and lower portions, moving to light volcanic pumice as aggregate for the dome.

Today's builders aren't able to draw on a slave workforce as the Romans could. In addition, the economic basis of modern construction precludes the use of some of their labor-intensive techniques. Engineers, however, have developed a successful roller-compacted concrete, now used by the U.S. Bureau of Reclamation to build dams. It utilizes some of the Roman components and construction techniques. David Moore, an engineer who spent part of his career at the bureau on dam construction, pioneered the study of Roman concrete and its implications for modern construction.

Furthermore, modern engineers have been able to duplicate the miracle ingredient employed by the Romans. Instead of procuring volcanic ash, the industry is using fly ash, blast furnace slag, and other materials. Formerly discarded as waste products to burden landfills, these materials can now be recycled into productive new life as Roman-style concrete. Another promising substitute for pozzolanic ash: rice hulls. It's hoped that the ingenious reuse of all these substances will also reduce the serious amount of CO_2 or greenhouse gas produced by the manufacture of concrete.

How important is all this? It's pretty significant. Every single year, a shocking amount of concrete gets produced. There are more than six billion folks on the planet, and your share of concrete production is roughly one cubic yard of the stuff. Creating a wealth of superconcrete out of such waste products is a huge boost toward a cleaner, safer, sturdier planet. Other structures, from bridges to high-rises, may now have the good fortune to stand tall for millennia, just as the Pantheon, marvelous temple to all the gods of its day, continues to do.

THIRST FOR THE GOOD LIFE

Rome's first aqueduct was built by a guy nicknamed Appius the Blind. An attempt at Roman humor? Possibly. Some accounts say that Appius Claudius Caecus did lose his sight later in life, the cause attributed to a god's displeasure.

Appius liked to stir things up. A blueblood himself, he opened the door for the lower classes to hold political office when he took the first of several high offices and became a Roman censor. His choices of plebians and freedmen for senatorial office ticked everyone off, including his co-censor, who quit. This allowed Appius Claudius Caecus to put his own architectural stamp on two projects of major import: Italy's first real highway, immediately dubbed "the queen of roads," and Rome's first effort to bring fresh water to the city. On both he boldly put his name: they're called the Appian Way and the Aqua Appia.

A whirlwind of activity, Appius accomplished these feats in the space of five years, beginning in 312 B.C. Rome wasn't very urban yet, but local demand had already polluted springs, wells, and the Tiber River. Thus the Aqua Appia, which delivered abundant good water from springs ten miles east of Rome, was a huge public works accomplishment.

Like the other aqueducts that would eventually serve Rome and grace the Italian countryside, most of the structure ran underground. Only 5 percent of the aqueduct used arches or graceful arcades of arches to carry its channels, and then only because they were needed to maintain the proper gradient for gravity to keep the water flowing. (Later aqueducts that incorporated more arches, such as the A.D. 52 Aqua Claudia, encountered frequent problems with maintenance and repair.)

Nearly everyone has heard about Rome's marvelous water delivery system, one that supplied the ancient city at its peak population of one million with a gazillion gallons of fresh water daily. (Quarrels over the actual amount supplied, and whether or not it exceeds that supplied each day to New York City, continue to smolder.) Most of Rome's aqueduct system supplied facilities that were open to the public: baths, neighborhood fountains, public latrines. But Rome was just the beginning. From Appius' time, and more especially during the imperial centuries, engineers and architects fanned out around the expanding empire to design and build aqueducts and water delivery systems appropriate to a variety of terrains and sites.

When it came to aqueduct beauty, utility, and longevity, Roman architects routinely outdid themselves. Example? These pristine arcades in Segovia, Spain.

In places such as France at Nemausus (present-day Nimes), in Spain at Segovia and Tarragona, and in Asia Minor at Smyrna (Izmir, Turkey) they stacked triple rows of arches to make arcades, on top of which sat roofed concrete channels where the water flowed. These picturesque structures, with their huge mellow stones, still span gorges and valleys today.

Engineers employed other ingenious methods when confronted with difficult, up-and-down terrain. Using gravity-pressurized pipelines called inverted siphons, they were able to convey water up slopes without too much loss of flow and successfully solved the issues of airlock, sediment formation, and atmospheric pressure differences. At Pergamum in Asia Minor, an affluent city since Hellenistic times, its water delivery system overcame severe vertical challenges in that fashion. So did the cities of Aspendos in Asia Minor, Lugdunum (Lyons today) in Gaul, and Lincoln in Britain.

You may be wondering why such a volume of fresh water was even needed, given that many cities throughout the Roman Empire sat on rivers or lakes. The answer? Lifestyle. Call it the liquid Roman dream. As Rome conquered regions and established colonies, civic planners laid out new cities and even permanent military installations according to a pedestrian-oriented grid that called for civic buildings, temples, and public baths.

Baths large and small became the biggest gobblers of water. Although aqueducts began to appear prior to the advent of the larger baths called thermae, in later years the demand for huge quantities of water grew, due to the popularity of daily bathing at all social levels. The convenience and health benefits of plentiful drinking water, urban irrigation, and continuous-flow toilets were add-on benefits.

Even vanished cities could be brought back to life in Roman style. Carthage, for instance. A superpower and Rome's formidable rival, it had been defeated and then utterly razed to the ground in 146 B.C. The Romans salted the earth to ensure the city would not rise again. Nearly three centuries later, however, Emperor Hadrian visited and decided to establish a new, improved Carthage. To supply its needs, the phoenix city got the longest and most sophisticated aqueduct ever built in ancient times. When finished, its water flowed at an estimated 98 gallons per second. In what is now Tunisia, the aqueduct sprawls nearly 82 miles from its mountain source, with some of its 65-foot-high pillars and its twenty-four quonset-hut-shaped cisterns for water storage still intact.

If he were alive today, Appius Claudius Caecus might gape with astonishment at the beautiful evidence in stone of what he began twenty-eight hundred years ago. As he himself said in a famous speech: "Every man is the architect of his own fortune."

THE BIG CHILL B.C.

Perhaps you assumed that iced drinks and designer waters, the two leading brands of which now soak up $1.6 billion annually in the United States alone, were modern inventions. Not so, at least around the Mediterranean in ancient times. Cities and towns from Corinth to Capua vied with one another as to who had the sweetest springs and the tastiest water source.

In its aquatic heyday, Romans raved about the liquid delivered by the Aqua Marcia aqueduct, which rose at the end of the Paelignian Mountains, calling it the coolest and most wholesome anywhere. Fans of the water from the Aqueduct Virgo, however, insisted theirs was even cooler.

Since ease of handling and convenience weren't considered virtues, water was lugged in heavy amphorae from local fountains to homes, then decanted into unglazed pottery, where osmosis and evaporation chilled the contents further. (In the early twentieth century, German researchers Von Luschan and Dollinger experimented with thin-walled porous pottery, achieving results that chilled the contents below room temperature by as much as 41 to 70 degrees Fahrenheit. Who needs refrigeration with that kind of cool?)

Then as now, both the elite sipper and the tippler in the street relished even more bracing liquids, cooled with snow or ice. As a result, snow shops in big urban areas such as Rome and Antioch did a lively business. Wealthy households used a fancy metal strainer to filter wine through the snow into individual cups. Humbler drinkers got the same result by filtering their wine coolers through special coarse cloth.

From the sixth century B.C. forward, households employed a variety of earthenware serving vessels to chill liquids. One useful device was the double-walled amphora, whose outer chamber could be filled with snow to give drinks a glacial edge or with hot water for mulled wines. Other pottery shapes included floating containers for snow or ice and stand-alone vessels with outflow openings near the bottom.

The secret to refrigerating or freezing food, along with producing snow and ice in quantity without Freon or electricity, was labor-intensive but fairly low-tech. Entrepreneurs took donkey trains into the mountains of Italy, Greece, Lebanon, or Turkey, harvested snow, then stored it in large underground pits or caves lined with chaff or straw. Pits were made deep enough so that the weight and pressure of the snow on top gradually turned the bottom layer to hard ice. Sometimes seasonal, high-end foodstuffs were brought to the snow regions for quick freezing—for example, asparagus, an imperial favorite. To haul snow or ice back to the city, workers simply covered it with rough cloth that acted as insulation.

An otherwise undistinguished teenage emperor named Elagabalus once cooled his summer palace with snow—Rome's first attempt at AC.

Way back in 1700 B.C., underground icehouses had become standard add-ons for palatial digs in red-hot lands such as the upper Euphrates (modern Syria). For instance, the rulers of the wealthy kingdom of Mari, Zimri-Lim and his wife, Shibtu, possessed an icehouse, an amenity that came in handy to chill the pomegranate-flavored beer and red wine coolers that Mari royals favored.

In the fourth century B.C., Alexander the Great was rumored to have borrowed such icehouse technology from cultures farther east. While besieging the city of Petras (in modern-day Jordan) he had thirty ice pits built and covered with oak boughs. As even a child knew back then, sieges gave attacking generals and officers a powerful thirst for chilled wine.

Even prior to this, ordinary folks on hot dry Greek islands had been making underground refrigerators each summer to chill their water jars.

Ice could be harvested from high-altitude ponds, but most of it was manufactured using the snow-pit process. It was icier and pricier than snow. Paralleling our processed bottled water today, frozen water in any form cost more than the vino it chilled. Pliny the Younger once wrote a crabby letter on that very issue to a no-show guest, along with an ultimatum: "The snow [for the wine at dinner] I'll most certainly bill you for and at a high rate—it was ruined in the serving."

Crabbiness at the other extreme came from the Roman philosopher Seneca, food crank and chronic sufferer of gastrointestinal ailments. He was anything but philosophical about icy delights: "Nothing is cold enough for some people—now and then you'll see them throw lumps of snow into their cups."

He wasn't alone. Medicos from Hippocrates through Galen harrumphed about the use of snow and ice to cool drinks, but by and large they were ignored—at least by the healthy. Several doctors flourished by employing a contrarian stance. One of these was the physician who attended Octavian Augustus, Rome's first emperor. He "cured" the emperor of liver problems, perhaps by frightening the malady away, since his solution involved a strict regimen of nothing but ice water to drink, lettuce to eat, and baths taken in frigid waters.

Although the Mediterranean climate remains as delightful as it was then, the city of Rome could roast for days on end. Private villas had fountains, other water features inside and out, and servants who wet down floors and worked fans, but that was about it. Curiously enough, Rome's most wasteful and worthless emperor, the Syrian-born teen Elagabalus, may have come up with the first methodical attempt at cooling the air rather than the drinks. One summer, he had huge quantities of snow brought from the mountains and packed around his pleasure garden. Possibly more of an igloo effect than air-conditioning, but it did the job. Needless to say, no one had the means to follow his profligate ways—or the courage to one-up the emperor.

ARCHITECTURAL TRIUMPHS, AND TURN-ONS

Here's a living history puzzle the whole family can do. Take a road trip around Spain, France, Italy, Sicily, North Africa, Turkey, and Syria; throw in a side trip to Croatia and Bulgaria, noting all those spectacular Roman bridges, arches, aqueducts, and arenas you pass. Wonder why they're still upright?

Quality building and materials, some would say. True, but an overarching reason might be the arch, an item the Romans borrowed from the Egyptians or the Etruscans. To give the Romans their due, they brought the arch to a sublime level of utility and beauty.

The darned things proved useful everywhere. Need a hydrodynamic shape for a cloaca sewer? Try an arch, or a series of them. A rambunctious river to cross? Use a series of arched spans to support the flat part of your bridge and you won't regret it.

True arches came in multiple flavors: the cusped arch, the basket-handle arch, the stilted arch. Your everyday engineer didn't have much truck with these effete innovations. Give him a good stout arch with a keystone at the top, and he was happy. The V-shaped stone or brick at the highest point of the arch, the keystone took the weight from above and distributed it in equal fashion down both sides. That allowed builders more flexibility to build up as well as down.

Arches with keystones could also support more weight than earlier designs, allowing for bigger structures. To begin, workers erected a semicircular wooden framework, over which the stones or bricks of the arch were assembled. Only when the keystone was locked into place did workers take down the timbers.

In A.D. *315, Roman emperor Constantine dedicated his triumphal arch, decorating much of it with artwork from prior administrations.*

The arcade was another superb idea. Besides looking beautiful and spanning great heights, the mass of an arcade—a horizontal string of arches—maintained the integrity of each section. With the weight evenly distributed, architects could build arcades without fear of collapse.

Standard arches did have their drawbacks. To avoid becoming fallen arches, the legs or piers had to be stout, since the sides had a tendency to bulge outward. When used to support bridges, the semicircular shape of arches presented a challenge, sometimes requiring the roadway to rise in the middle. This type of construction also demanded terrific carpenters, since semicircular supports of wood had to be constructed under each arch being built.

Long before the Romans, the Greeks had used the post-and-lintel system to create their classical signature look—the boxy Greek temple. Now, freed from the constraints of rectangularity by the arch, the Romans embraced the semicircle and the dome, soaring to new heights with their buildings.

Having examined the Cloaca Maxima, the sewer system put into place by the earlier Etruscans, engineers had already spotted how the arch could be extended as a series, creating barrel vaults. Inspired by this out-of-the-box thinking, Romans tried intersecting a pair of barrel vaults and came up with a highly stable variation called the groin vault. Emperors exulted: now we're talking! Henceforth, engineers could erect wonders like the Baths of Caracalla, with its airy, light-filled interiors and vast roofed spaces.

Even structures with more utilitarian purposes, such as flat-roofed warehouses, could gladden the eye. In Rome, a stunning example was the Portico of Aemilia, which integrated six side-by-side rows of barrel vaults on three stepped terraces, giving clerestory lighting to the interior of the building.

Not every project could be as hugely juicy as the previous examples. But arch builders also cashed in on an early fad that grew in grandiosity and narcissism as time went on. Back in Roman republic days, posturing individuals could win permission to build a special arch or barrel vault dedicated to themselves. Such a monument to ego gratification looked similar to the arcus or triumphal arch but was called a fornix to differentiate them.

Located on pedestrian thoroughfares and forums, offering shade and a modicum of privacy, fornices became favored loitering spots. To the consternation of arch honorees, they also became primo business locations for Roman whores, whose unsavory activities took their name from the structure itself: fornication.

In contrast, the more genteel triumphal arches began as temporary structures, decorated wooden pass-throughs under which a victorious general rode in his chariot for his postwar triumph. As time went on, they morphed into more permanent structures. By A.D. 85, Rome boasted 36 triumphal arches; three of them still remain as picturesque examples. Such arches stood in high-traffic areas so everyone could admire the bas-relief artwork, showing in near-photographic detail the suffering inflicted on a particular people by the triumphee.

The arch of Titus showed his A.D. 70 campaign in Judea, including the loot, such as the menorah carried by soldiers in his triumph. By A.D. 312, the time of Emperor Constantine, his celebratory arch had swollen into a three-holer, fully loaded with columns, statues, and carvings. A thrifty sort, Constantine cut costs by pilfering statues and other elements from monuments to prior emperors around Rome.

In the Roman mind, bridges and arches went together like caveat and emptor. Not so for the Minoan engineers of ancient Crete, who'd built their long-ago bridges of stone slabs. They'd probably dreamed up the corbelled arch as well, a misnomer since it needed help to counteract the effects of gravity. It seems petty to carp, though, since a Minoan corbelled arch bridge still stands in Crete a mere thirty-nine hundred years after the ribbon cutting.

Speaking of the ability of bridges to stand the test of time, here's a dandy. An early wooden span over the Tiber River connecting to Tiber Island went up in 192 B.C. Just 130 years later, a bureaucrat looked at the now-shoddy thing and said, "Upgrade it to stone." Bids went out, and one was accepted; the contractor got cracking. Up went a two-arched structure of peperino limestone, complete with an ingenious flood hole in one pier to deal with high-water periods on the Tiber. The final touch? A large inscription, stating that L. Fabricius, *curator viarum* or supervisor of roads, had overseen the construction, then inspected and approved the structure.

Hallelujah—payday dead ahead for the bridge builder, you'd assume. You'd be wrong. The Romans, cynically wise to the ways of the building world, made contractors guarantee their work for forty years. Not until the forty-first year did the builders recoup their initial investment!

ALL'S UNFAIR IN LOVE AND WAR

The term "right-hand man" might have been coined to describe Marcus Vipsanius Agrippa, the faithful second in command to Octavian Augustus, Rome's first emperor.

Every inch the military commander, Marcus Agrippa also excelled at peacetime accomplishments.

Of humble birth, Marcus signed up as a youngster to serve in Julius Caesar's army. His combination of brains, good sense, and way with the gladius sword made such an impression that Caesar sent him to study with his own teenage grand-nephew Octavian at a swanky philosophical school in Apollonia. (Apollonia, then an upscale city on the Ionian Sea, has since become an unremarkable landlocked burg in Albania.)

The two teens became friends. That's where they were when the shattering news of Caesar's assassination arrived. There were decades of battles, both political and literal, before Octavian took sole power as the second Caesar. Agrippa became his not-so-secret weapon. A shrewd, farsighted general as well as admiral, he outmaneuvered volatile Marc Antony and myriad other opponents on land and sea.

Agrippa gave twenty-three years of steadfast military and political support to his new emperor. That identification, coupled with his frequent portrayals in films, have made later generations feel they know him. His engineering achievements, however, were even more substantial—and surprising to most of us.

The man was a builder. More than that, he generously subordinated his own genius at city planning and architectural projects to make his commander's feats shine with greater glory. The building program for Rome that the two childhood friends put into motion was long-term and spectacular, designed to wow the crowd and impress the old boys. Its only parallel might have been the Pericles-driven "golden decades" of building on Athens' Acropolis—but Rome's transformation was citywide, not merely at the sacred sites.

In 34 and 33 B.C., Marcus Agrippa embarked on a huge public works program that he paid for himself, rebuilding aqueducts and repairing infrastructure. He'd never forgotten his plebian roots and completed dozens of projects—from public baths to parks to voting places for the plebians—that contributed mightily to the health, welfare, and pleasure of Rome's common people. (Parts of the Baths of Agrippa complex can still be seen in Rome; so can the Pantheon, which still says "Made by Agrippa" over its entrance. Although the smaller original that he built burned down, Emperor Hadrian erected a stunning successor on its foundation.)

The work Marcus did drove his pal Octavian's popularity rating through the roof. In a few short years, they would need that public support. In 31 B.C., the two men faced a final pair of lethal and wealthy opponents—Egypt's Queen Cleopatra VII and the renegade Marc Antony. At the famous sea battle of Actium, Octavian took his friend Agrippa's advice and they won a massive victory—both physically and psychologically.

Once he finally became first man in Rome in 27 B.C., Octavian Augustus pushed Agrippa to serve in several public offices in Rome during his career, including consul, praetor, and aedile. What Marcus Agrippa most relished, however, was being a part-time general and planner-builder without borders. He headed up construction of the first major road network linking Rome with its provinces and cities. It spiderwebbed from Italy into Gaul and Spain, from the Atlantic Ocean to the Rhine River.

Highway mission accomplished, Agrippa trotted off to Gaul and Spain to do other wide-ranging projects. He founded the city of Caesaraugusta (today known as Zaragoza), built the splendid theater complex at Merida, Spain, and constructed a naval base at Cadiz. In southern France, there are arches, bridges, temples, and other structures attributed to him that still stand in Nimes, Lyons, and elsewere. One standout: the slender beauty of the Maison Carré temple.

This indefatigable man waged war wherever he was sent to do so, then built things in the peaceful aftermath. Places from Antioch to Athens saw his handiwork, and he founded colonies in Greece, Asia Minor, and Lebanon.

In his laughable leisure time, Marcus wrote on geography. The emperor also gave him the task of measuring the exact size of each province in the Roman Empire, which he did by recording the data from the milestones along all the imperial highways he had overseen. This survey of point-to-point distances took Marcus twenty years to complete—after which he had the master map mounted on the wall of the Portico of Vipsania in Rome.

During the forty-one years of Octavian Augustus' reign, that map would be copied countless times for military brass and administration officials. A special copy was engraved onto a circular piece of marble for the emperor himself.

Ironically enough, the foremost geographers and encyclopedists of the day had access to the raw data but didn't understand its true value. Even Pliny, who copied some of Agrippa's figures for distances, failed to make much use of it. As a result, the geographical portions of Pliny's *Natural History* are its weakest.

Was there anything that Marcus feared to tackle? Maybe one thing. A rational man, an engineer at heart, he held the conventional religious beliefs of his time—but worried quite a bit about the darker side of human nature. While aedile of Rome in 33 B.C., he carried out a PR campaign to expel all astrologers and sorcerers from the city. In some ways, it appears, he feared for his old friend's safety. On more than one occasion, he warned Octavian Augustus about the harm that magicians, foreign religious elements, and fake philosophers could perpetrate.

History played a terrible trick on poor Marcus Vipsanius Agrippa. He'd always been the good guy, done what Octavian asked him to do—had even gotten married and divorced on command. Of the three wives Octavian foisted on him, the last was the emperor's own slut of a daughter Julia. Together Marcus and Julia begat five children, which led to a grandson better left unbegotten: Gaius Caesar, better known as Caligula. Now *that* was a fearsome legacy. If only Agrippa could have stuck to mapmaking.

THE WASTELAND

No one is dead certain where the Etruscans came from; some historians believe they hailed from Asia Minor or had a long-ago link with the Minoans of Crete. Like the Minoans, they were fastidious when it came to waste management.

The later Romans were the beneficiaries of their mania for civic tidying up. Once they elbowed the Etruscans out of "their" new capital, they found a sewer system already in place. The losers claimed it had been installed by the Etruscan king Tarquin, who did so after a flood had made a soupy mess of the low ground where he fancied building a forum, or maybe a racetrack.

Called the Cloaca Maxima, the sewer system was an ingenious network of arches, vaults, and tunnels. Part of it consisted of uncovered canals or channels at street level, where residents could dump the contents of their chamber pots and cesspools. Belowground, massive channels allowed wastewater and unspeakable solids to move beneath the city, discharging it all into the waters of the Tiber River. Even when Rome was still small and dingy, the Tiber was beginning to look—and smell—pretty dubious.

Tarquin's Cloaca ran from the forum to the Tiber River. The main line was made of volcanic tufa from Brocchi, cut and dry-set tightly into three concentric arches, 13 feet wide and high enough to permit small boats to sail through it. The strength of these triple-vaulted ceilings was sufficient to keep them intact for millennia, until the present day.

Now and then, city rulers would repair conduits or make additions, but it was not until the forty-one-year reign of Rome's first emperor, Octavian Augustus, that a major overhaul of the Cloaca took place. The emperor had long counted on the willingness of Marcus Agrippa, his childhood friend and powerful right-hand man, to lead in battle and fight for him politically. This time he really pushed the envelope,

Agrippa even tackled Rome's aging sewer system, the Cloaca Maxima. Onlookers called it the Labors of Hercules: the Sequel.

sending Agrippa where no sane man chose to go: the wasteland below the city of Rome.

By now the admirable Marcus thought he must be channeling Hercules, given the Herculean tasks his dear old school chum asked of him. And Hercules only had to contend with animal dung. With a huge sigh, Agrippa began swamping out the stenchy kingdom below Rome. He was a bit over-scheduled, what with massive building programs aboveground, political duties, and so forth, but between 33 and 15 B.C. he managed to squeeze in his noisome new assignment.

After surveying the whole disgusting maze by boat, he ordered extensions and additions to the lines; ultimately, the single main became three. New vaults and arches were built with hard peperino limestone from the Gabine hills, some with concrete faced with bricks. In places, Marcus decided that the floor of the Cloaca actually needed revamping. Hundreds of smaller drains, together with feeder lines of lead or clay pipes to elite residences and neighborhoods, were added or upgraded and cleaned. Like the fatty arteries of Roman gourmands, over time the walls of the Cloaca had become clogged as the flowing waters deposited calcium carbonate on them, and required periodic grinding.

Where the aqueducts joined the sewer system, there were covered catch basins for sediment settling. At that point, the aqueduct water got distributed through canals and pipes to reservoirs and then to users. Most end users—except for the very wealthy and the cleverly criminal who bribed water officials—were not connected directly to fresh water, but had to fetch it from the neighborhood fountains that flowed nonstop throughout Rome.

After aqueduct water made its way through the public baths and fountains, it then picked up the dreck of countless latrines and small industries, carrying its final load of raw sewage, corpses, gray water, and heavy metal grunge to the Tiber, there to make its diseased way to the sea.

Mired in the purgatory of endless sewer redos, Agrippa found that he also needed to add hundreds of neighborhood fountains to the system, along with paving these areas and much more besides. Oh yes, and the actual course of the Tiber River itself required a new embankment—plus a new bridge in the same vicinity. It's sad but not entirely surprising to learn that Marcus Vipsanius Agrippa died in 12 B.C., most likely worn out by his germ-laden labors for Rome. He was fifty-one.

FATAL SPECTACLES ON FAKE LAKES

How to impress the public while keeping death row populations to a minimum? A challenge first met by Julius Caesar, whose showmanship in every arena, including literal arenas, galvanized everyone. In 46 B.C., to celebrate his brutal wins over the Gauls and others, multiple triumphs were held in his honor. By Jove, that didn't feel special enough to Julius, so he ordered his legions of legionaries, who were all standing around anyway, to break ground in Rome's Field of Mars. A trillion shovelfuls later, a manmade lake. Upon its waters, Julius organized the first naumachia, or mock sea battle, the Romans had ever witnessed.

His sham naval conflict involved rival fleets of trireme warships manned by four thousand rowers and two thousand fighters, costumed as Egyptian warriors, versus the bad Phoenician dudes from Tyre. In one critical sense, this reenactment of a famous battle wasn't really "mock" at all, since the "actors" were doomed to perish. With the foresight he was famed for, Caesar used POWs, captured enemy soldiers, as the cast for his watery inaugural.

Romans weren't easy to awe, but mock naval battles scored high in imaginative carnage.

Was his naumachia a hit? Julius thought so. The crowds were huge; several unlucky fans were crushed to death, including two Roman senators. Three years later, after Caesar's assassination, city officials had to fill in the fake lake, which was now smelling to high heaven.

His prototype spectacle would be imitated by a lineup of Roman emperors. They would, however, use condemned criminals for the most part, filling the gaps with POWs and a few cameo roles played by paid gladiators.

Caesar's aqueous spectacle eventually goosed his adopted son Octavian, later known as Emperor Octavian Augustus, to mount his own naumachia. Determined to both honor and outdo his now-deified predecessor, Augustus dug a much bigger giant hole, over 2 million square feet, on the west side of the Tiber River. To fill it, he commissioned a new 12-mile-long aqueduct, the Alsietina.

In the middle of his imperial lake, Augustus put an island with a sturdy bridge to it, allowing apparatus for the spectacle to be moved about easily. He reasoned that an island offered room for creativity, and Roman crowds loved novelty: hungry crocs in the water, fake volcanoes, scenes from *The Odyssey*, men dressed as cannibals fighting unlikely foes.

Built with tiered seating, the Naumachia Augusti was surrounded by brick walls, shady groves, and gardens. The structure's gates had 6-foot-deep canals leading to them, easily accommodating the 121-foot triremes, whose shallow draft was about 3 feet. The empty vessels, their oars stowed upright, were hauled by crews with ropes through the canals. (Given the number of ships in play, it's likely there were one-way "enter" and "exit" canals.)

The emperor assembled between three thousand and six thousand doomed men to reenact the Battle of Salamis, where Athenians had whipped the Persian navy. The audiences ate up the action on the 50-acre lake. This day in the year 2 B.C. coincided with the grand opening of the temple of Mars Ultor, aka "the Avenger." Augustus built it to remind Rome of his victory over the assassins of his adopted father. A few decades tardy, but he'd had construction issues; that darned aqueduct took forever.

The lake built by Augustus was later used by Emperors Nero and Titus. Possibly it was restored in A.D. 247 by Philip the Arab, Rome's most obscure emperor, who wanted to make a big splash for the millennial anniversary of the city's founding. The trouble with spectacles was, the next one had to be more spectacular. Otherwise, what was the point? Emperors stressed mightily over this conundrum.

After A.D. 80, a few naumachiae put on by Titus and Domitian were supposedly held in Rome's spiffy Colosseum. After a remodel that added an underground labyrinth of chambers, however, the arena could no longer be flooded for sea battles. Some historians are doubtful that the Colosseum ever could have held a naumachia, given the difficulties of filling and emptying it of water in a reasonable amount of time.

Being extravaganzas, naumachiae were held infrequently. Most were colossal hits—but when they bombed, they bombed big time. During the tenure of Emperor Claudius, that carnage-mad, strangely lovable sexagenarian, he wanted to flaunt his tremendous engineering feat: an underground channel to drain Fucine Lake for flood protection. Running from the mountains to the lowlands, it took thirty thousand men more than a decade to complete.

Thus instead of a dug-to-order pond, his water fight took place in a lake-filled valley. On it, a hundred warships, filled with nineteen thousand armed combatants, maneuvered in a reenactment of a sea battle between the Rhodians and the Sicilians. A huge audience crowded onto the slopes around the lake.

Claudius had thought of everything. To keep the vast prisoner navy of desperados from escaping, soldiers armed with swords and catapults were stationed on rafts around the lake's perimeter.

Once the sea battle concluded, the fighters having shed blood and shown bravery, a number were pardoned. Claudius then gave the signal to open the waterway of the channel in order to drain the lake, so that a gladiator free-for-all could be fought on semidry land. To everyone's horror, it instantly became clear that channel and lake weren't properly aligned. A flood of Noah-like proportions began to spurt and spew, sweeping away combatants, spectators, and dignitaries alike.

Claudius and and his wife, Agrippina, were spared drowning. Instead, they conducted a shouting match about illicit profits and whose fault it was. Agrippina grew particularly incensed at the water spots on her one-of-a-kind gold-embroidered cloak. The spectators who survived, however, had great stories to tell their grandchildren about this soaking-wet fiasco.

For sheer disaster, meteorological and human, the Claudius washout was topped five emperors later by Domitian, who built yet another artificial pond alongside the Tiber, surrounding it with a magnificent arcaded stone structure, with seating for thousands. His naval slaughter unfolded as planned, its prisoner cast of thousands turning the waters to gore.

Suddenly a violent storm struck. Domitian, however, gave orders that no one could leave. As icy rain hammered down, the emperor changed into a thick woolen cloak but forbade others to put on dry clothing (not that most people had brought any). As a consequence, many people fell ill and died of exposure. One of the strangest sadists ever to wear the purple, Domitian then made the most peculiar amends. As author Dio Cassius described it, "By way, no doubt, of consoling the people, he provided at public expense a dinner lasting all night."

MONUMENT TO THE UNKNOWN ARCHITECT

His name is a mystery. Yet nineteen hundred years ago, it must have been on everyone's lips. Of all the impossible jobs a Roman architect could be asked to do, his was the worst.

First problem was the site, if you could call it that. For centuries, it had had a swampy stream running through it. As Rome got more crowded, slums were built on it. After the Great Fire, the area was dug out and made into a

Christians plus lions plus Colosseum were perennial favorites with moviemakers. It's doubtful, however, that martyrs were even nibbled on here.

lake for Emperor Nero. Along its shores sprawled Nero's Golden House, a palatial jumble of architectural hubris.

All that changed in late A.D. 69 when Vespasian, first of the Flavian emperors, delivered the assignment from Hades.

"I want to put my stamp on the city," he said to the unknown architect. "Something splashy that obliterates the awful memory of Nero. Instead of letting the heart of Rome be monopolized for an emperor's private pleasure, let's give the people a palace of pleasure. Build me our first real amphitheater—a big one that'll hold fifty thousand people or so."

Once the architect was revived after his faint, he set to work. There were no precedents in Rome. Aside from two small arenas that had been destroyed in the A.D. 64 fire, all previous venues for gladiator events had been temporary.

The mind of the unknown architect swirled with numbers. His project consisted of two jobs: drain and fill in the lake, then build the amphitheater. Emperor Vespasian wanted big? He'd give him big. The architect envisioned a huge oval, four arcaded stories high. Lousy with travertine marble, eighty entranceways.

After the ordeal of draining the lake, he got bad news. To keep the underpinnings of the amphitheater dry, to build the foundation, and to install the drainage network would occupy years and would gobble up labor, materials, and money. Lots of money.

Fortunately, finances weren't a problem. Emperor Vespasian and his son Titus had returned from having crushed the Jewish rebellion with endless booty, displaying some of it during their triumph in A.D. 71.

The architect labored on, his growing workforce composed partly of Rome's poorest men, a pragmatic solution that addressed the city's urgently high unemployment rate. Unlike prior emperors, Vespasian sprang from a middle-class family. A good-humored, tolerant man of the people, yet cosmopolitan, too, he was an emperor who'd spent years in the military from Britain to Asia Minor, and as an official from North Africa to Greece and Thrace.

Numbers, numbers, numbers. The architect had already raised the valley floor with 13 feet of fill; now he raised the entire site another 20 feet or so—only then to begin the devilish task of digging out the foundation, which would go down nearly 40 feet under the walls and seating. Altogether, he oversaw the movement of more than 220,000 tons of earth and rock from the gigantic hole, all of it removed by oxcarts and the hands of grunt labor.

The foundation itself? A massive base of concrete-and-rubble perimeter walls 10 feet thick and 39 feet high, the center then filled with a mixture of concrete, mortar, lime, sand, volcanic rock, and water. (Modern estimates: sitework and foundation work alone would cost over $50 million.)

The fun stuff at last, the now ulcer-ridden architect told himself. Not so fast—there were still the drains. Like New Orleans, the site of the amphitheater sat very low, wanting nothing more than to become a lake again. When it rained, the building would act like a huge cistern. And what about public latrines and the water needed to wash away blood and animal waste? He solved the problem with more math, more labor, more hydraulics to supply flow from the Claudia aqueduct and channel wastewater off into the Tiber River.

For the aboveground structure itself, his solution was elegantly simple. He used the ideal 5-to-3 ratio of arena length to width, fudging it slightly to accommodate his master plan for 80 external arches. The arena ended up being 615 feet long and 510 feet wide.

Little by little, his elliptical masterpiece rose into the cobalt sky of Rome, the ground floor adorned with three-quarter columns in the Dorian order; the second story, Ionic; the third story, delicate Corinthian columns. Strong concrete-cored arches connected to make a series of dramatic arcades lacing around the structure.

Some of the most wondrous parts of the architect's scheme were hidden, and meant to be so. The vomitoria, for example—the clever maze of stairways and passages that enabled fifty thousand spectators to get in or out of the building in fifteen minutes. (All seats were assigned tickets, and by rank; women in the nosebleed seats at the top, senators and celebrities in the front-row marble chairs, and the rest in between.)

In June of A.D. 79, the unknown architect finished the third story of his amphitheater and got shocking news: his sixty-nine-year-old employer,

Vespasian, was dead. His son Titus, a sensible and likeable go-getter, smoothly took over. Two months later, however, Emperor Titus had his hands full with relief efforts. Mt. Vesuvius had erupted and southern Italy lay in ruins, entire cities now hidden deeper in the earth than the foundations of the amphitheater.

Soon Titus began to urge the architect to hurry up and finish. After such a national tragedy, it had become more important than ever to give people a distraction. Titus planned to honor his dead father—and the Flavian dynasty—at the grand opening of the amphitheater.

The home stretch. By now, after ten years on the job, the unknown architect felt like an old man. In A.D. 80, he saw the very last piece set into place—a bronze plaque that identified Emperor Vespasian as the "architect" of the new amphitheater and the source of its funding. It made no mention of the real architect.

The festivities began. The biggest structure ever built in the Roman Empire glowed like glory on its site. Inside its concentric rings, the first of one hundred days of events took place, a long parade of death in which nine thousand animals and unspecified numbers of humans gave their lives before a packed house.

The amphitheater, officially named after the Flavian dynasty, would become Rome's symbol, but by a different name: the Colosseum. Emperors Vespasian and Titus had tried but could not wipe the slate of collective memory clean. Somehow, Nero's name, attached to the colossal statue he'd had made of himself, clung indelibly to the building. And does so, even now.

None of this probably mattered to the unknown architect, whose name is the one we long to know.

FRANCHISING BLOOD AND SAND

The first permanent home for the gladiatorial games wasn't in Rome at all. That amphitheater sat in southern Italy near Neapolis (Naples), the pride of the little city of Capua, which considered itself the heart of the gladiatorial training schools.

Some years later, around 80 B.C., a stone arena that seated nearly twenty thousand rabid fans went up in Pompeii. "Up" may be an inexact word to describe its architectural style. In the early days of gladiator spectacles, promoters took advantage of natural hollows in the landscape, banking up earth to make them into ad hoc arenas. Although the Pompeii amphitheater was built of wood and stonework, it sat in a similar bowl. To enter, fans climbed external steps to the top of the facility.

This became an ill-fated venue—and not just for the men and animals that fought and died in its arena. In A.D. 59, during the reign of Emperor Nero, a dazzling gladiatorial event was promised by a dubious fellow who'd been chucked out of the Roman Senate. For his splashy inaugural, it appears he hadn't hired enough security—and furthermore, he didn't know enough to pat down ticket holders for illegal weapons.

At the show, fans from the nearby town of Nuceria got into an ugly squabble with Pompeiian fans. Stones began to fly, swords were drawn, and at the end of the brawl, hundreds lay dead, with many more wounded. After a senatorial investigation, the people of Pompeii were declared culpable. As punishment, their amphitheater was shut down for ten years. That really hurt.

The earlier Etruscans, who originally controlled the region of Campania, imparted their way of honoring freshly dead notables with one-on-one gladiatorial smackdowns, called *munera* (meaning "service"). The Campanians

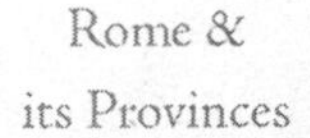

Other cities demanded versions of the Colosseum. Arenas in North Africa eventually showcased many Christian martyrs, with and without lions.

gleefully took them up. Later the munera became known as ludi or funeral games. As the notion expanded, it lost most of its religious connotations but added buckets of blood and ever more elaborate events, thanks to sponsors with deep pockets and political ambitions.

As demand for the ludi grew, they were held in temporary sites such as the Roman Forum or the cattle marketplace, which simply required temporary wooden grandstands. That sufficed from 264 B.C. on. Once Julius Caesar, on his way to becoming political top dog, stepped up to sponsor the first really lavish gorefest of 320 pairs of gladiators, the idea of a permanent facility gained traction, but nothing happened until 29 B.C., when a small wood-and-stonework amphitheater was built in the Field of Mars by the Taurus family. Inexplicably, nobody liked it much. In A.D. 57, Emperor Nero, ever eager to pander to the public, built another wooden amphitheater. Before long, it and the Taurus arena burned in the Great Fire of A.D. 64.

In A.D. 80, after a twelve-year wait, the architecturally spectacular, fifty-thousand-seat Colosseum was inaugurated to rave reviews. Only then did the hysteria for arenas begin to resonate around the Roman Empire. Just a handful of cities already had permanent amphitheaters: Merida in Spain, Lyons in France, Carthage in North Africa. But every city, up-and-coming town, and colony founded by one emperor or another already possessed the template of Roman urban planning: public baths, straight-as-a-die street grids, and aqueducts. What they all screamed for now was a permanent place from which to view blood-and-sand entertainment.

They got their wish. A few modest arenas were constructed against city walls, but most followed the architectural plan of the original Colosseum.

Despite their ignoble purpose, the Colosseum and its smaller clones possessed an architectural plan that was noble as well as sturdy and practical. The key elements: perfect elliptical shapes in a 5-to-3 ratio, heavy use of keystone arches in series and as stacked arcades, the entire structure supported by yards-thick foundations of concrete.

By the third century A.D., more than 230 towns and cities had amphitheaters to boast about. Arena addiction stretched from Albania to Cyprus, Tunisia to Wales, Israel to Switzerland. Because of the permanent presence of Roman troops in imperial times, a dozen or more were built for bored soldiers in Britain, Scotland, and Wales—a few repurposed from other structures.

In France more than thirty arenas were erected. Two are still in use; the sturdy, two-story one at Nimes, maintaining its gladiatorial aura with films, artwork, and demonstrations, and the amphitheater at Arles, retaining its bloody past by acting as a part-time venue for Spanish bullfights.

Some of these Colosseum also-rans still evoke awe—the stirring amphitheater at El Djem in Tunisia, for example. Now an archaeological site about two-thirds the size of Rome's Colosseum, it once seated thirty-five thousand people. Other arenas have been recycled into benign and useful purposes. The beautiful, airy amphitheater at Pula, Croatia, became part of an art installation in 2003, having a giant red cravat (apparently a national obsession) wrapped around it. But the best swords-into-plowshares usage may be that of the arena at Verona, Italy. This structure, predating the Colosseum, has seats of ancient pale pink stone that now support the bottoms of music lovers at bravura open-air performances of *Aïda*, *Carmen*, and other operas.

SEALSKIN—DON'T LEAVE HOME WITHOUT IT

The Romans were lightning freaks—astonished by the phenomenon and terrified by it. Since Jupiter, originally a sky god who handled all electrical matters, was their main deity, they did their utmost, collectively and individually, to placate him. And protect themselves—or think they did.

For instance, it was an article of faith that wearing laurel wreaths and sealskin protected humans from lightning strikes. Rome's first emperor, Octavian Augustus, never went anywhere without his sealskin wrap. He had good reason to be spooked: thunderstorms seemed to follow him around. When he first began to build a house for his family on the hills of Rome, lightning struck the site with great vigor. Straightaway the emperor vowed to build a temple on the still-smoking ground—then fearlessly built his home next door.

Ten years later, Augustus himself was nearly hit by a bolt from the sky god; instead, it smoked his litter bearer. This time, he set about building an even fancier temple to Jupiter Tonans, meaning "the thunderer," on Capitoline Hill.

In more-is-better fashion, Rome accumulated at least six storm deities and corresponding temples to house them. Besides Jupiter Tonans, there was Jupiter Fulgur, the daytime lightning god, and Summanus, who handled the night shift. Fulgora, the goddess of lightning, also was represented, as was Minerva (Athena in Greek), goddess of storms and credited with having invented the thunderbolt. None of this assiduous worship appeared to result in fewer storms and strikes, however.

In certain years, Rome really got pounded meteorologically. For example, in 65 B.C. lightning wrecked parts of the inner city, also zinging the bronze she-wolf statue, symbol of Rome. Other bolts completely melted the official table of laws. Asked for an interpretation, the augur priests warned that the city and the laws of Rome would soon founder. They were right—sort of; the Roman Republic was on its way out, although bigger changes would not occur until Julius Caesar took power.

Given the number of severe storms in Italy, the art of lightning interpretation became a critically important augury, addressed by diviners called fulgurators. They were aided by ancient tomes called Books of Lightning that had been written over the centuries, or so legend had it, and kept on hand for such occasions. According to them, lightning emanated from the planets Saturn, Jupiter, and Mars. Other experts put forth theories as to how it all worked; one asserted that storms came into being when two clouds began clashing. Since we now know that lightning is produced when a discharge of static electricity jumps from one cloud to another, that explanation wasn't far off the mark.

The one person who got it right in a scientific sense was that acute thinker Lucretius. In his first-century-B.C. epic poem *De rerum natura,* he described lightning as a fundamental force in the universe, a type of "rarified fire, made of minute and mobile particles to which absolutely nothing can bar the way."

Most fulgurators ignored such rationality; after all, they had a business to run, storms to predict, omens to read. To interpret lightning their way, the right way, practitioners divided the sky into sixteen regions, each run by a different divinity. (Nine were different aspects of Jupiter.) Lightning that came from the east meant good fortune; modern meteorologists would tend to agree, since that direction indicates the probable end of a storm. From the west was unlucky; again, that is the direction from which thunderstorms normally approach. To the Romans, the most dreaded omen was lightning from the northwest.

Fulgurators also distinguished between three sorts of lightning. The first kind merely charred the sites it hit. The second, a quick-penetrating variety, tended to wreck building interiors but leave exteriors largely undamaged. The third bad boys, accompanied by high-decibel thunder, destroyed everything.

Even if they could not control or predict the lightning, Rome's fulgurators played an important role. Before every assembly and official meeting, they scrutinized the heavens and took the auspices. If they declared that lightning was on the way, all meetings were canceled. If the gathering was already in progress, it was dismissed.

When a site got hit, it became sacred ground. The fulgurators cleaned it, buried the damage, fenced it in, and posted a sign that

Zeus, CEO of the Greek Olympian gods and as Jupiter, a feared favorite of the Romans, packed heat in the form of thunderbolts.

it was now the sole property of Jupiter. Sometimes a purifying sacrifice of a two-year-old sheep took place.

Lightning bolts sometimes left calling cards, called fulgurites, especially in sandy soil. The heat of the bolt would melt the available silica, making a glassy structure, sometimes as hollow tubes, root shapes, or lightning-shaped "fingers." Regarded as holy or as good omens, some were huge, like the sixth-century-B.C. fulgurite found at the Oracle of Dodona, a bronze copy of which can be seen in the Athens Museum of Archaeology.

On the other hand, if a person were unfortunate enough to be struck and killed by lightning, it was regarded as very unlucky for the larger community. The victim and family might have been pretty teed off as well—especially since he or she would be denied burial in the family crypt due to Jupiter's evident wrath. Instead, the victim was hastily buried at the site of his or her inadvertent electrocution.

Thunder gods were a touchy bunch. Around 277 B.C., a storm hit the temple of Jupiter Best and Greatest and lightning melted the head of a statue of Summanus, the nighttime deity. The fulgurators took this to mean that Summanus was demanding his very own temple. Plans promptly got under way to build one near the Circus Maximus, and the structure did fine until 197 B.C., when—sigh—it too was hit by lightning.

CHANNELING EPICURUS AND COMPANY

He had one of those tedious three-part names Roman aristocrats favored, but he was as sensible and forthright as any nature-loving country boy. Despite living in Italy during a time of bloody civil wars, slave revolts, assassinations, and demagoguery, young Titus Lucretius Carus accomplished the nearly impossible. He became the most eloquent champion of the earlier Epicurean school, clarifying and preserving the beliefs of Epicurus and the fifth-century-B.C. atomists Leucippus and Democritus. Furthermore, he did this by writing seventy-four hundred lines that addressed evolution, the workings of the atom, and much more—as an epic poem.

In his epic, which filled six books, this scientific bard also speculated about those so-called atoms as being agents of disease. No one else in the Greco-Roman world at the time talked about such crazy ideas. Everyone knew that illness came from an imbalance of the four humours, or perhaps an evil eye attack from an envious source. But sickness because of invisible bits of matter? What nonsense.

It's probable that Lucretius was a Roman born around 90 B.C., most likely a patrician whose patron was Gaius Memmius, to whom he dedicated his epic poem *De rerum natura*. Nevertheless, his biographical facts remain scanty.

In general, Romans cared much more about practical matters and applied technology than they did about philosophy or theories on the nature of the universe. Latin itself lacked an adequate vocabulary to discuss these matters. One of Lucretius' achievements was to forge a new vocabulary for Latin, beginning with the title of his book. *De rerum natura*, often translated as *On the Nature of Things*, more precisely meant *On the Coming into Being of Things in the World*.

The Greeks originally called this field of study *physika*, which to them meant "to come into being." Latin had no equivalent; thus Lucretius used the Latin verb "to be born" to derive his word *natura*. This in turn is how we arrived at the oversimplified English translation of *natura* as "nature."

By championing vital earlier works on science and philosophy, Roman poets Vergil and Lucretius saved them from extinction.

The young poet had no way to work directly with his chosen teacher, Epicurus, who'd died some 250 years earlier. Fortunately, his books were still available in Lucretius' time. In addition, around the Bay of Naples in southern Italy, numerous Epicureans lived and studied. Their main focal point? The philosopher Philodemus, who had an active circle of adherents in Herculaneum. (One of the delights unearthed in Pompeiian archaeology is the tattered, charred remains of the actual library of Philodemus. The hope is that more pieces of Lucretius' work will be found among the papyri.)

Given the decades-long political chaos throughout Italy and the general unease provoked by a string of ruthless politicos, dictators, and military commanders, people from all walks of life were perpetually terrified. Random acts of violence occurred everywhere. One of Lucretius' tasks, as he saw it, was to heal that anxiety and free humans from the fear of their own mortality through Epicurean teachings. An idea to be applauded, since such possibilities were extremely probable, given the climate.

With *De rerum natura*, he achieved much more than that. Lucretius left to posterity many great treasures of early scientific thought in addition to his clear and faithful rendering of the Epicurean system. Since the voluminous original works of Epicurus would be lost once the Roman Empire began to deteriorate, this indeed was a key legacy, and a selfless one.

Like Epicurus, Lucretius took a few scientific wrong turns. For example, their belief in the infallibility of sense perception led both men to declare that the moon and the sun were as tiny as they appeared in the sky.

He only lived some forty years; it's thought that the last book of *De rerum natura* may not be in the final form he envisioned. Nevertheless, famed writer-orator Cicero and his brother Quintus saw a probable late draft in February of A.D. 54 and declared that it contained many flashes of genius.

Lucretius died on the same day that another young Roman poet stood in the Italian city of Cremona (near Milan) for his manhood ceremony, wearing his first *toga virilis* at age seventeen. That poet was Vergil, and he would later honor Lucretius by writing, "Happy is he who has discovered the causes of things and has cast beneath his feet all fears, unavoidable fate, and the din of the devouring underworld."

Later Christians labeled Lucretius an enemy of religion, although he (and Epicurus, Democritus, and others) were more interested in ridding the world of superstition and the supernatural beliefs surrounding religion. Four centuries after Lucretius, the Christian writer Jerome would also seek to smear Lucretius' name with a tale worthy of today's top media sleazemongers. Jerome said that Lucretius had been driven mad by a love potion, wrote his epic during a few lucid intervals, then committed suicide, leaving behind an incomplete work that had been heavily edited by Cicero. Sex, madness, and plagiarism, all in one sound bite.

But Lucretius lives on, his work a towering influence on thinkers across a wide range of disciplines, from Darwin to Pierre Teilhard de Chardin. A most quotable philosopher, his words on superstition and science still ring true: "No fact is so obvious that it does not at first produce wonder; nor so wonderful that it does not eventually yield to belief."

DO AS I WRITE, NOT AS I DO

Roman Stoics believed ardently in apathy. Well, not ardently, since the whole idea was to avoid passions of any kind. Using apathy as a mental border guard, they would stop all emotions before they crossed the frontier of the soul. To writer-philosopher Lucius Annaeus Seneca, that was the recipe for happiness. Oops, scratch happiness—far too extreme a feeling. In Seneca's words of warning: "Delight is the mind's propulsion to weakness . . . to be in transports of delight is the melting away of virtue."

He was born about 3 B.C., most likely in his Roman Spanish family's hometown of Cordoba. The wealthy elder Seneca, a well-regarded figure in rhetoric, pushed his son in the same direction. As a youngster Seneca was tubercular, possibly asthmatic, and well on his way to becoming a career hypochondriac. After milking the ill health thing, he eventually gave in to Dad's nagging to jump on the standard Roman political track as orator-lawyer, writer, and officeholder.

His sharp oratory and edgy style, called "silver Latin" by the cognoscenti, won him fame. His lean good looks won him a few extramarital hookups with randy royals—and punitive action as well; in his thirties he was exiled by Emperor Caligula, and later by Emperor Claudius, to the island of Corsica. Plenty of time to practice that Stoic indifference to pain, rethink those high-spirited royal hookups, and study philosophy to help the years of exile go a bit faster.

Finally, after strenuous lobbying by Emperor Claudius' niece Agrippina, Seneca (now pushing fifty) got recalled to Rome as a teacher for a preteen Nero. Besides the plushness and prestige of the post, he got to pontificate away, displaying his erudition, his turns of phrase, his wide knowledge of matters philosophical.

The only worm in the apple: Nero himself. After the sudden but not very mysterious death of Emperor Claudius, the pimply seventeen-year-old became emperor. Seneca, now chief speechwriter, became the behind-the-scenes minister who called the shots. Stockpiling wealth, snubbing senators, he lived the high life for five years or so, until the skyrocketing mortality rate among Nero's family members and inner circle became hard to ignore.

As if that downturn weren't enough to deal with, less affluent Stoics started calling him a hypocrite. Seneca had a standard reply: he'd resigned himself in true Stoic fashion to whatever fate sent. Okay, so he was a multimillionaire; he would have to bear it uncomplainingly.

By A.D. 63, now a filthy rich, terribly scrawny senior citizen and far from political favor, Seneca tried to worm back into the imperial graces. He wrote fulsome pieces on the golden age of Nero. He signed over his immense real estate holdings to the emperor. He wrote a book on natural science that would become a source of misinformation for centuries. He retired from public life. Despite his low profile, in A.D. 66 he was invited to commit suicide after the discovery of another imperial conspiracy.

Even that didn't go well. Seneca tried to slit his wrists, but not enough blood emerged. He then drank hemlock that evidently was past the sell-by date. Finally he got into a red-hot steam bath and managed to suffocate.

The disparity between the Stoic Seneca's words and his not-so-stoic acts was aptly remarked upon centuries later by Thomas Carlyle: "Notable Seneca, so wistfully desirous to stand well with Truth, and yet not ill with Nero . . . the niceliest-proportioned Half-and-half, the plausiblest Plausible on record; no great man, no true man, no man at all."

Bossed by the twin terrors of Nero and his mother, Seneca wrote what they wanted to hear. Toadying failed to pay off.

Postmortem, Seneca took on a weird new identity. The growing Christian community glommed on to the man they saw as a kindred spirit. Granted, the actual Seneca had written a number of thoughtful essays—one on the cruelty of the gladiatorial arena, another on the brotherly treatment of slaves—that did argue for a humanitarian, Golden Rule sort of approach. But this adulation went a few hundred laps beyond the facts. Someone concocted a whole correspondence between Seneca and Paul, who were indeed contemporaries. In these letters, Seneca acts as the kindly bug in Nero's ear, trying to spring Paul from house arrest. He also reads some of Paul's writings to the emperor in an effort to win him over to Christianity. Another enthusiast created the notion that one of Seneca's literary tragedies (tragically lost, of course) was the story of Christ's cruxificion, which in turn must have influenced the writing of the Gospel narratives, especially those of Mark and Luke.

These and other fictional deeds got Seneca into the catalog of saints put together by Jerome, as well as other warm accolades from later Christian writers such as Tertullian. The would-be Stoic philosopher might also find it ironic that in this fashion, he has become an Internet immortal.

VERY HAIRY PORTENTS

Even before it had a name, Halley's comet was making drive-by visits to planet Earth. In 467 B.C., a banner year for splashy celestial bodies, Halley's was spotted on one side of the world by a keen-eyed Ionian Greek named Anaxagoras and confirmed on the other by Chinese astronomer Ho Peng Yoke. The comet's return engagements at roughly sixty-seven-year intervals were also noted at least eight times, although no one in East or West cared to name it and claim it.

When an object as eerie as a naked-eye comet showed up, it got attention—particularly since it often hung around for days or weeks. People of long ago got seriously creeped out by the sparkling hairy visitors, considering them portents of evil events: the death of a ruler, the start of a war, the hint of an earthquake. Since natural disasters, wars, and administration changes were common events, they often coincided with comet appearances. Or reports to that effect. As the historian Livy sagely remarked in 218 B.C.: "In Rome or in the neighborhood of the city many prodigies occurred that winter, or rather, as is wont to happen when men's mind are roused toward matters of faith, many were reported and too easily believed."

No artificial lighting dimmed the awful grandeur of comets; small wonder they were looked on as messengers of doom.

Natural scientists of the day took stabs at understanding comets but were thwarted by their often oddball beliefs about what constituted "the sky," what the objects in it were, and what—if anything—held them up.

Aristotle posited that as the sun heated the earth, it gave off a flammable vapor, which in turn caught fire from time to time, producing comets. He sneered at earlier thinkers such as Pythagoras, who thought there was only one comet, a planet that just showed up from time to time. Apollonius of Myndos, on the other hand, hypothesized that many comets had planet status, describing them as "bloody, menacing, carrying the omen of bloodshed to come."

In his circa 77 encyclopedia of fascinating facts, myths, and gross errors, the Roman writer Pliny organized comets (or hairy stars, as the Romans called them) according to their shape and appearance: torch, horn, javelin, discus, horse's mane, and bearded, among others.

Somewhat earlier in the first century A.D., Roman philosopher-author Seneca had sidled closer to the truth. In his book *Natural Questions*, he said that comets had closed orbits and were permanent voyagers around the heavens. He also urged further study of comets, advice which was not taken until the seventeenth century by Edmond Halley.

A variety of comets with dazzling tails did click with gloomy predictions or ghastly events. For example, a run of smaller comets was seen when Pompey the Great and Caesar were battling it out for supremacy between 56 and 48 B.C.

But the most famous and fortuitous comet was the Big Guy of 44 B.C. Lacking a tail, surrounded by shimmering rays, the thing was visible in broad daylight. Horrified Romans saw it show up during the July games thrown for the deceased Caesar by Julius's adopted son Octavia, the eventual ruler of Rome.

The young Caesar-to-be deftly turned a nasty omen into a PR plus, assuring everyone that the oddly star-like comet had arrived on cue to honor his predecessor and put a stamp of approval on his new administration. As he later wrote, "During my Games a comet was visible for seven days in the northern part of the sky. The common people believed this star signified the soul of Caesar received among the spirits of the immortal gods, and on this account the emblem of a star was added to the bust of Caesar that we shortly afterwards dedicated in the forum."

This century's historians and astronomers have also studied the track record of ancient comets. One of the most interesting findings: in A.D. 66, Halley's comet showed up again. The brightest and most luminous comet of ancient times, it arrived over Italy and lingered there a record nine and a half weeks to plague Nero, last of the Julio-Claudian emperors. By now the twenty-seven-year-old Emperor had killed an inordinate number of family members and civic leaders. He'd been implicated in Rome's Great Fire of A.D. 64, which he sleazily scapegoated onto the city's small community of Christians. When Halley's showed up, Nero demanded an explanation for this latest celestial event, preferably one that would keep him from being torn to bits by a fearful mob.

Philosopher-author Seneca, who'd been Nero's top advisor and mentor but had fallen from favor, found himself really up against it. He did a smooth job on the media release, calling it "the recent one during the reign of Nero Caesar—which has redeemed comets from their bad character." Nice try, but the ever more paranoid Nero wasn't buying it.

Despite his groveling in print and in person, Seneca was accused of conspiracy. He and others were forced to commit suicide, proving at the very least that comets could be grim news indeed for higher-ups in the Nero administration.

SWEETLY DAMNING EVIDENCE

Most lead compounds are toxic yet taste sweet, making lead-laden toys irresistible to toddlers, who regularly explore their world with hands and mouths. They've got good reasons: kids begin life with substantially more taste buds than adults and their tongue-tip receptors are more sensitive to sugars.

Two thousand years ago, Greek and Roman societies—and not just the underage crowd—found that sweet metallic flavor equally irresistible. While toddlers sucked on toys made of lead, their parents consumed a wide variety of lead-enriched wines, syrups, and sauces. This slow-motion toxification took place all around the Mediterranean, affecting Roman citizens and noncitizens, adults and unborn children alike.

Like us, ancient consumers had a sweet tooth; honey, their sole sweetener, was too scarce and costly to meet demand. Then some well-meaning idiot found that vinegar, poured onto lead, rendered the liquid sweeter. The result, called sugar of lead, seemed just the thing to counteract often-sour vin ordinaire, which was frequently slow-simmered with spices into a hot drink.

In a parallel development, Greek and Roman cooks learned to reduce unfermented grape juice to concentrate its natural sugars. Their culinary secret? Lead cookware. The resulting high-octane syrup (called *defrutum* or *sapa*) was added to fruit preserves and also to entrees from meat to fish—much as high-fructose syrup is added to countless processed foods today.

Long-ago winemaking procedures introduced additional perils. Inside fermenting vats, lead strips were glued to lids, then sealed for forty days. If the strips showed no signs of corrosion after opening, the wine was declared fit to drink. Smaller storage vessels were often lead-lined, too. The upper classes and wannabes regularly consumed their wine beverages from goblets of lead or pewter (a tin-lead alloy).

Given this sweetly damning evidence, the average Gaius of the first century A.D. may have gotten a substantial daily dose of lead just from the food and drink he consumed.

What about the lead pipes and holding tanks through which his daily water flowed? Author-architect Vitruvius, who lived and wrote in the first century B.C., had seen the effects of acute lead poisoning. As he wrote, "In plumbers, the natural color of the body is replaced by a deep pallor. For when lead is smelted in casting, the fumes from it settle upon their members, and day after day burn out and take away all the virtues of the blood from their limbs. Hence, water ought by no means to be conducted in lead pipes, if we want to have it wholesome."

Lead plumbing certainly wasn't good for anyone. But compared to the larger dangers presented by their contaminated wine and food, drinking water from lead plumbing was a lesser worry. Then, too, over time Rome's water delivery channels became caked with calcium carbonate deposits that actually protected users from the lead.

Sip by sip, nibble by nibble, countless unsuspecting consumers throughout the Roman Empire accumulated enough of a lead burden to experience the early symptoms of saturnism or lead poisoning: a persistent metallic taste and loss of appetite. No surprise, then, that strongly flavored sauces such as garum, a fermented fish liquid that became the Mediterranean world's catsup, grew widely popular with people from all walks of life.

To stimulate patrician taste buds, cookbook gourmands such as Apicius made rich dishes loaded with spices fashionable. One of his gravy recipes called for generous amounts of pepper, turmeric, garum, caraway, cumin, ginger, and asafoetida root.

Roman mouths got the worst of it in other ways, too. When decay caused a cavity, dental experts of the day usually opted to pull the tooth. To get a better grip, the tooth was "prepared" by pouring molten lead into the hole! Lead held the ailing molar together until it could be extracted—if the agonized sufferer hadn't committed suicide or homicide by then.

Women got even more exposure through the use of cosmetics, lead ointments as spermicides, and lead-rich compounds as abortifacients. Mothers would have been anguished had they realized the toxic effects on their offspring. Lead passes easily through the placenta, causing miscarriages, birth defects, mental retardation, and brain damage.

Why did lead become so ubiquitous? Simple. Lead was the plastic of Greco-Roman times. A common element, easily processed with a low melting point, malleable yet strong, useful alone or in combination with other elements, it seemed—as plastics once did—the "miracle" material.

Manufacturers and users had no way of testing its toxicity. Many Romans and Greeks knew that lead fumes could make people ill, but imbibing it in other ways caused no immediate alarm. Lead isn't excreted from the body but slowly accumulates in bones, like plutonium does. It injures stealthily, over time.

The processes of getting lead from ore and manufacturing with it loaded its workers with poisonous fumes and lead dust, making their lives indescribably awful—but most end users did not think about these matters too deeply. The harsh realities of the times dictated that workers in the extractive industries were usually slaves or criminals whose death sentence was to work until they dropped.

Archaeology has shown that Rome and the vast regions it ruled for centuries were far from primitive. Most people lived in a complex consumer society of mass-produced goods that turned towns and cities around the Mediterranean and Black Seas into smaller versions of Rome. What they did lack were consumer watchdogs and modern science to help them identify and assess environmental risks to themselves and their children.

There's an unsettling lesson in Rome's cascade of events and the role that lead may have played in its health and welfare. Instead of smirking at the lack of awareness in the ancient world, we'd do well to take a harder look at our own track record.

Lead-filled paint and leaded gasoline are now banned in this country, but it took decades and nasty, David-versus-Goliath fights. Recalls of toxic toys and other children's products have occurred on a regular basis and safety regs continue to be tightened in order to protect kids—but what of the millions of lead-contaminated toys and other gear already in households?

Equally troubling is the way we export our toxins in the form of junked electronics for poorer nations to deal with. Leaded gas is still being burned today in myriad countries from Morocco to North Korea.

LIPS TO DIE FOR

Drop-dead beauty—that's what the affluent Roman matron sought. Given the near-lethal cosmetics she used to get it, the drop-dead part was almost literal. Gone were the classic Greek days of demureness. Everyone wanted smooth pale skin, flashing eyes, and ruby lips. Since "everyone" in this instance consisted of a critical mass of married women or those about to be wed, all

Gimme another shot of that asses' milk—my face is starting to peel off again.

of them vocal when it came to looking good, a corresponding number of shysters, snake-oil salesmen, and manufacturers rose to the challenge.

In ancient times, cold cream was about the only beauty aid that didn't actively poison its users. The Greeks favored a krinos-lily unguent made in the flower-growing town of Chaeronea. Much later, the prolix Roman author Galen, that well-known sawbones about town, dreamed up a concoction of rosebuds steeped in a 1-to-3 ratio of wax and lanolin. To achieve the proper cooling effect, water was then incorporated into the scented mass. Also vying for cold cream queen was Poppaea Sabina, the notorious second wife of the emperor Nero, who invented a sleek body butter she named after herself.

Not long ago, researchers actually found a full jar of face cream in a Roman-era temple in London; it had been overlooked since the second century A.D. After chemical analysis, it was found to contain starch, tin oxide, and animal fats. Chemists opine that the makeup of the ancient goop was quite similar to the ingredients in modern face creams—although they shied away from guaranteeing that no animals were harmed in the manufacture of the eighteen-hundred-year-old beauty aid.

When it came to face pampering, Empress Poppaea was her own best advertisement. She also dreamed up a face mask made of honey, grated bread, rare oils, and scents, which she washed off each morning with the milk of an ass. Not just any old ass, either. Thanks to the obscene wealth of her Pompeiian family, a stable of five hundred female asses provided her daily milk bath.

Even middle-class women in Greece and elsewhere had a nightly beauty regimen featuring a mask of some sort, all of them guaranteed to remove freckles, blemishes, and wrinkles. A common ingredient? Bean meal. The beauty mask recommended by Dioscorides, the famed herbalist, combined rosin, rose extract, and Chian earth, which was a kaolin clay. Other lightening, tightening ingredients might include honey, egg white, and resins. Skin peels were popular, too: just apply white lead and sublimate of mercury and watch two ugly layers of epidermis slough off! Thanks to ample borrowing from early Egyptian chemists, Greeks and Romans also knew how to mix animal fat with lead salts to make lead soap.

Nocturnal beauty treatments were just the noxious beginning. Onto clean faces, Greek and Roman women buffed quantities of white face powder. The most popular—and deadly—was made of pure lead carbonate (toxic left-overs have been found at various archaeological sites). For exclusivity's sake, fashion leaders among the Roman elite demanded an expensive powder made from the white excrement of crocodiles. Or, if croc supplies ran low, just a dusting of arsenic. White chalk and orris root also served as face powder and were safer, too, although most women took a pass because, you know, they just didn't cling that well.

Both lips and cheeks got generous applications of rouge. The base mix included harmless ingredients such as mulberry, lichen, alkanet root, and sea-weed. In order to achieve that vibrantly desirable scarlet color, however, nothing could compare with minium, raw cinnabar, or vermillion. A type of red lead, minium these days is usually found in batteries and rustproof paint. Cinnabar, known as red sulfide of mercury, contained 86 percent mercury and 13 percent sulfur; vermillion was refined from raw cinnabar. Both came from Almaden, Spain, site of the world's most productive mercury mine for two thousand years—and just as toxically productive today.

Reading this litany of dire ingredients, it's easy to feel superior to those wretched fashion-conscious women of two thousand years ago, forced to choose between toxins and ghastly pale lips. The ugly truth? The issue is still with us.

A 2007 product test by the Commission for Safe Cosmetics revealed that one-third of U.S.-made bright red lipsticks tested contained unacceptable levels of lead; more than half had detectable levels. None listed lead as an ingredient. Like cinnabar, lead doesn't belong on lips or in stomachs. Over a lifetime of daily lipstick wearing, a woman swallows an amazing amount of it—four pounds, on average.

The eye makeup used long ago presented similar problems, just different toxins. Galena, the natural ore of lead, was the standard eyeliner. So was kohl, which until very recently was thought to be protective or at least benign. The Romans and Greeks also picked up the Egyptian fondness for green malachite, made from copper ore. The one apparently harmless product was the use of golden saffron as eye shadow.

Personal slaves and women of humbler station would seem to have dodged the toxic bullet when it came to Greco-Roman cosmetics. Not entirely, however. There was a long-standing belief that human saliva was an admirable ingredient in cosmetics of all sorts. Thus female slaves were obliged to chew certain cosmetics before applying them to their mistresses.

Beauty regimens—they've always been hell. And hell on the health, too.

WIDE-OPEN TO INTERPRETATION

You couldn't fault the Greeks for trying to perceive the future, or at the very least divine what the dickens the gods wanted—via divination.

Early in their history, they tried divination using animals. They kept an eagle eye, so to speak, on the movements of the eagle, symbol of Zeus, chief of the Olympian gods. They noted every move made by owls, the totemic mascot of Athena.

Greeks felt more at home, however, getting advice from oracles, especially ones with an inspired intermediary like the Pythia at Delphi, who gave her forecasts in enigmatic verse. The Greeks loved her predictions, repeating them for centuries. For instance, in 550 B.C., King Croesus of Lydia, who always topped the list of the world's richest men, asked the Pythia if he should pursue his military campaign against the Persians. She replied, "Do so, and you will destroy a great empire." Thrilled, he went ahead with his plans. The O. Henry twist: the empire that took it in the shorts was his own.

Combining a look at the future with a nice outing, oracle pilgrims journeyed to a variety of divination sites, most of them shrines of enduring fame. One favorite lay in northern Greece. At Dodona, sacred to Zeus and filled with his emblematic oak groves, the divination news came "by trees," as it was called. Priests listened to the rustling of oak leaves and foliage and delivered instant messages. If an oracle seeker preferred predictions "by bronze," the diviner interpreted the moans and whistles of the wind around the bronze statues and bowls at the shrine.

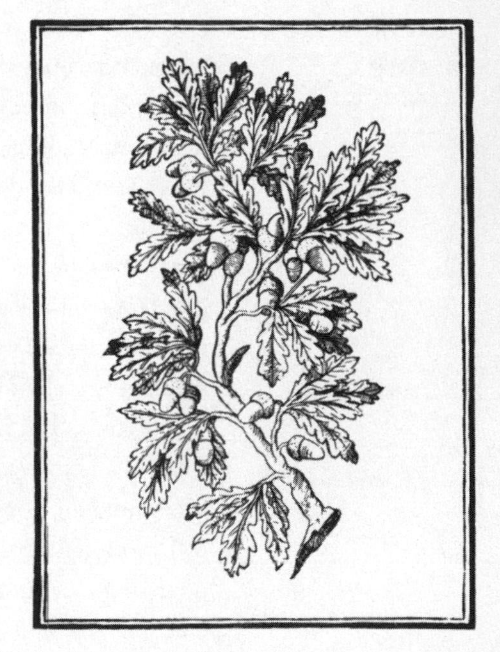

At the oracle of Dodona in northern Greece, the leaves of the oak, sacred to Zeus, provided instant message divination.

Initially the Greeks felt quite content with the array of prediction mechanisms they had. Ultimately, however, it took the Romans and their obsession for detail and hairsplitting to expand the product line of the whole divination industry into a splendid Baskin-Robbins array of choices that quickly spread around the Med.

The Romans were latecomers who'd come in and squashed the local Etruscans in order to dominate their territories north of the city. Later they borrowed an

Etruscan cultural crumb or two—such as entrail reading, called haruspicy.

For home altars, sacrifices to one deity or another could be made with milk or wine or plain salt cakes. More momentous sacrifices involved animals: doves, rabbits, dogs, pigs, sheep, and cattle. Actual sacrifice—making the offering sacred by barbecuing it—was the final step.

Prior to that, the animal had to be inspected to ensure it was flawless and worthy of the god; only then would it be autopsied. After assistants killed and opened up the victim, the haruspex or diviner plunged his hand into the animal's carcass and removed several vital organs. To the trained eye, the organs, principally the liver and gallbladder, were prediction road maps. To learn the ropes, trainees worked with scale-model bronze livers that showed every lobe, lump, and curve of the organ. The bronze model, divided into sixteen sectors, showed the domains of forty gods and other significant factors.

Entrail reading, although splashy in multiple senses of the word, and a calling that employed a great many people, was in close rivalry with another group of religious professionals: the augurs. Chosen from the patrician class in Rome, augurs were organized into colleges. Augury (the word originally meant "the study of bird flight, birdsong, and other avian stuff") took place on the summit of Capitoline Hill. Since birds were the messengers of Jupiter, to them augurs went to get an auspicious answer. Not just any crummy starling would do, however: the desired birds for augury were the eagle, vulture, owl, raven, and crow. To get under way, the augur marked out a portion of sky, pitched a tent, and sat there waiting for favorable signs. It was like an on-location film shoot. Everyone nearby had to keep quiet—no sneezing, coughing, or other sound effects, or the whole thing was declared invalid. Once the augur had spotted an auspicious sign, he called it. Inauspicious signs could clear a public assembly or scotch a planned battle in minutes.

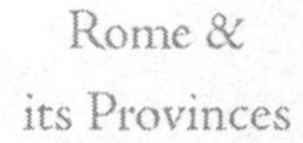

Nine out of ten generals used divination—including Julius Caesar, who threw dice for advice before crossing the Rubicon River.

More everyday divination utilized a dizzying array of methods. Hydromancy involved reading the patterns of oil poured on water. Ignispicium looked for clues in ignited matter. Another favorite was divination using mirrors. If a mirror wasn't handy, the shiny surface of a military shield would suffice.

Fans of alectryomancy preferred divination using some sort of fowl. The alphabet was traced in sand, then sprinkled with wheat or barley. As the chicken pecked the letters, meanings were sought in the words—a literal gobbledegook sort of method.

Perhaps the most popular divination choice was cleromancy or the drawing of lots, using dice, knucklebones, or dried beans. When Julius Caesar finished demolishing Gaul and headed for Rome to challenge Pompey, he halted before crossing the Rubicon River, the official border of the Roman state. On its muddy banks, he did a quick cleromancy, then declared, "The die is cast!"

A much later variation of cleromancy cooked up in the third century A.D., called the lots of Astrampsychus, had the questioner choose one of ninety-two set questions, then do some math tricks to arrive at a number in a table of answers from oracular gods. By this date, there also was a Christian version of this divination, using saints' names.

Weather and natural phenomena provided nice outdoorsy ways to divine and predict. Lightning activity, storms, cloud formations, stars, and celestial movements all had their adherents. The howling of dogs or canine actions out of the ordinary were also studied for hidden meaning.

In short, almost anything, anyone, or any action could be used during the heyday of Greco-Roman divination. The three strangest? Teratoskopos, the interpretation of human deformities; divination by examining moles and birthmarks; and my personal favorite, twitch divination.

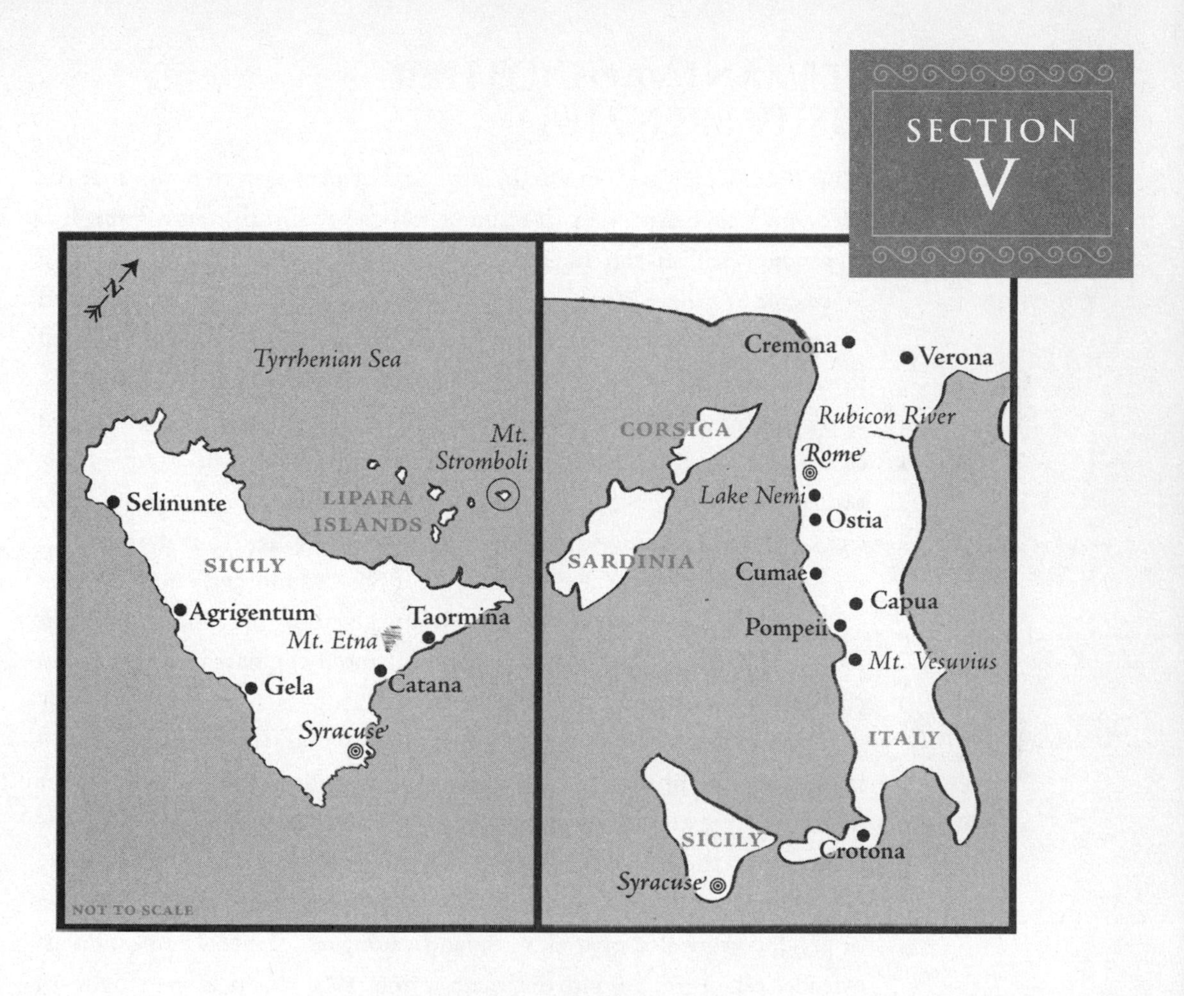

Italy & Sicily

EELMANIA AND OTHER FISHY FRENZIES

More than twenty-two hundred years ago, Italians began to notice that the Mediterranean Sea (never that resource-rich to begin with) was running low on fish and shellfish. You think today's prices for halibut and salmon are steep? You wouldn't believe what was paid for mullet, a delectable fish that weighed two pounds or less. As one long-ago writer gloomily put it, "Now the price of three horses is given for a cook, and the price of three cooks for a fish."

Except for those shore dwellers who made their living from the sea or lived close enough to drop in a line, fresh seafood, particularly salt-water species, was largely the province of the rich. The gift of a whole fish was considered a sign of the giver's lavish generosity—or his insane desire to buy friends.

As a consequence, the more abundant flesh of morays, lampreys, and conger eels became a gourmet item, swooned over by rich foodies and their parasitical friends. Paeans were composed about the flavor: "Tastier than chicken—seriously!"

During the two centuries leading up to imperial rule, the old patrician families and members of the senatorial class struggled to outdo one another in terms of prestigious display. One answer was seashore real estate: view villas around the Bay of Naples and along the southern coast, the larger the better. Someone introduced an ingenious fillip—the glorious artificial piscina or pond, surrounded by walkways, landscaping, and fantasy arches. Ponds extended estate grounds into the water, at times their channels open to the sea but gridded to keep the fishes captive. Aquaculture, Italian style, was born.

Piscinae spread faster than algae bloom. In sheer numbers and centuries-long staying power, aquaculture looked like big business. That's just what it wasn't, however. Piscinae of this era were more like edible yachts: items into which serious money was poured in a continuous stream. The superbly fresh, tasty creatures raised for the dinner table often cost owners more to feed and house than they would fetch in the market. Some specimens had to be fed on live fish of smaller species. Others required a diet of bird droppings, insisted experts.

Eels, on the other hand, had less exacting requirements: they got fat on fruit, milk curds, dried figs, and seafood that had gone off. They gobbled up other items, too—at times even the occasional slave thrown to them "as an example to staff."

The mechanically complex infrastructure for the entire setup was labor-intensive, including fishermen to trap and restock the ponds. Proud owners became inordinately fond of their captives, often relating to them as pets. Guests were dragged outdoors to meet finny individuals, some of which lived for decades. Dinner parties were held poolside so that diners could enjoy and envy the charming aquatic goings-on. The species called gray mullet supposedly had keen hearing and would come when called.

Eels, although largely raised for banquet fare, became pets as well. These snaggle-toothed, hideous family members at times outlived their owners. Eel freaks like celebrated orator Hortensius wept when their aged eels expired.

But the years passed and things changed. As the extravagant "my piscina's more Disney than yours" one-upmanship of the Roman Republic morphed into imperial governance, the ruling class of senators and other noble families lost privileges and powers. In time, a succession of emperors co-opted the exclusive right to build vast fishponds. And, as emperors were wont to do, through bequests, gifts, and skullduggery, they managed to scarf up an inordinate number of the finest seaside estates as well.

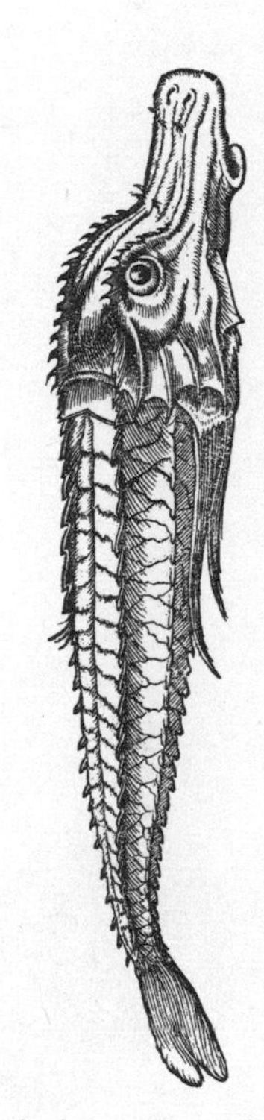

The lamprey, gushed over as a homely pet and raved about as a cheaper seafood source.

Around the same time, big social and technological changes were taking place. Water systems in Rome and southern Italy's Campania district got four new aqueducts, making more fresh water available for fishponds. The widespread use of hydraulic concrete made construction of underwater structures cheaper and easier. That technology, combined with growing numbers of socially ambitious families of knightly rank and freedmen who'd been successful in business, led to a new boom in more modest piscinae, mostly freshwater ones. The template for pond and fish farming operations, no longer confined to Italy, spread widely around the Med.

This new breed of fish farmers and eel growers meant business, often using techniques learned from earlier fish cognoscenti. Their piscinae had two to ten tanks, allowing various species to be raised. An ingenious method of increasing yield exploited the sex appeal of female mullets. Individual fish would be released into the sea with fishing line through their gills, then reeled in once they had attracted suitors of the opposite sex. Owners also raised fast-breeding tilapia, kept in check with hungry African catfish.

Although mullet, sea bass, turbot, bream, sole, and parrotfish remained perennial favorites, some piscinae operations also cultivated oysters, mussels, eels, and water snails. The last, an acquired taste, were massive mollusks raised on special diets, at times laced with wine.

There were two dandy pluses about raising eels. First, most species lived equally well in salt, brackish, or fresh water, untroubled by temperature extremes. Second, even though morays and certain other varieties ballooned to a chunky 40 pounds or more and grew to over 6 feet long, the fishes did not mind crowding.

Giant snails, anyone? Water-loving species were farmed and eaten with gusto by Roman gourmets.

Unlike the patricians, who grew faint at the thought of engaging directly in business (although they seldom kicked up a fuss if middlemen did their piscatorial dirty work), the nouveau riche piscina owners were pleased to benefit from aquaculture technology and prosper by harvesting their own fruits of the sea.

The climax of the cult of piscinae came in the Roman Empire's ripe old age. That was when artificial fishponds and oyster farms reached their tarted-up finest moments—as tourist attractions. As Mercury, god of travelers and hucksters, knew, where there's an attraction, there's an opportunity to sell a gaudy and tasteless souvenir. The intriguing finds of archaeology now include small glass flasks that date to the third and fourth centuries A.D. On them, etched in busy if not immediately understandable profusion, are the ponds, piscinae, and other must-sees of Baie and other coastal towns—along with cheery phrases along the lines of: "Very happy memories, friend."

IT WAS FASCINATION, I KNOW . . .

Considering the grave dangers that men, women, and children of long ago actually faced, it's a piquant fact that the issue they may have stressed over the most was fascination. We're talking about the evil eye, called *oculus facinus* or "fascinating eye" in Latin. Writer Horace called it *obiquus oculus*, "the unlucky glance that poisons." (Belief in the evil eye is still alive and well in Italy, where it's now called by the less poetic term *mal'occhio*.)

Worse than the evil eye or the Medusa, the basilisk had a glare that could kill at a distance.

The Italians weren't alone in this conviction. The most universal and persistent ocular superstition is the belief that the eye is capable of projecting and inflicting an injury on humans, animals, or whatever its unlucky gaze happens to fall upon. Furthermore, there is the equally strong certainty—still quite alive in the world today—that humans, animals, and even inanimate objects suffer from evil-eye rays.

Anthropologists and psychologists call this notion "an institutional recognition of envy." Others point to the fact that worldwide, many of those accused of having the evil eye have tended to be societal outcasts, marginal people (women living on their own and clearly up to no good), or outsiders, especially traditional enemies.

Besides old biddies bent on terrorizing brides and infants, numerous other parties in ancient times were suspected of evil-eye capabilities. Damned furriners often got tagged with the evil-eye label. Every right-thinking Roman knew that Greeks from Thebes, Cyprus, and Crete were well endowed with the ability to send literally withering glances. Not only did the tribes and peoples around the Black Sea, from the mysterious Thibii of Pontus to Scythian women, possess it, but several historians hysterically asserted that such individuals had two pupils in each eye for greater malignancy. Illyrians and certain exotic tribes in Africa were culprits, according to Greek and Roman writers from Herodotus to Pliny. Individuals with blue eyes, a squint, or a mole near the iris—all highly suspicious.

Who were most susceptible to being fascinated? First and foremost, baby animals and human children. Other high-risk targets were adults during moments of high happiness or good fortune—brides, for instance, and generals on the day of their military triumph.

Keeping in mind this preoccupation, one anecdote about Julius Caesar, who once celebrated four back-to-back triumphs for his military victories, makes more sense. Triumphs were daylong affairs, costing the earth and attended by everyone. During the long processions, to absorb the poison of malign stares at Julius from the crowd, Secret Service–level measures were taken. Protective amulets, in particular the apotropaic phallus emblem taken from Caesar's own hearth, dangled from his chariot, along with other amulets secreted on his person.

Thousands of Caesar's troops marched in the triumph as well. To further divert sinister rays from their commander, the men sang salacious songs full of derogatory remarks about Julius. They called him "the bald adulterer," and husbands were warned to guard their wives whenever he was in town. (As it happened, these digs were quite true in Caesar's case.)

That protective measure of denigrating the object is still practiced today, as when we say, "Oh, this old thing?" upon being complimented on an outfit. Among the Romans and Greeks, a more everyday method of neutralizing evil eye encounters was to spit three times on one's chest. Another option? Making the *mano fico* sign, the closed fist with a thumb through it, which allowed men and women to ward off negative glances while going about their daily business.

Evil-eye protection began at birth. People from all walks of life wore amulets, the minimal keep-safe action against fascination. These talismans could take various forms: complex snarls of threads, images of frogs or crickets (a Greek favorite), horn shapes, and the *mano fico* sign. Representations of the phallus, called the *fascinum*, were the odds-on favorite for anti-fascinating. Some phalluses bore eyes or wings, which made them more potent and vigilant protectors. Babies were decked out with a variety of such symbols. The favored teething ring for infants was a phallus made of coral.

Since homes and workplaces needed protection, Romans and Greeks placed the fascinum everywhere. That lively abundance of anatomically correct phalluses as well as the supersized ones in the countless paintings and statues of the god Priapus used to puzzle archaeologists. The sheer quantity of exuberant male organs found in gardens, at entrances, imbedded in mosaics, carved on pavements, and emblazoned over the doors of bakeries and workshops still embarrasses visitors to Italy, who tend to read a lot more into the sex lives of the ancients than was probably true.

These days, archaeological museums have storage drawers overflowing with fascinum objects and protective amulets, some of them exquisite works of art. But what particular havoc was it thought that the evil eye could wreak?

The list began with the withering of crops; one of the earliest laws enacted by Romans forbid the fascination or enchantment of the fields. The evil eye could also destroy domestic animals, kill babies, make adults ill, or give them eye diseases. It also could bring misfortune to the recipient, from broken arms to broken engagements—and is still thought to do so today in many cultures worldwide.

Other eyes besides human ones could also bring ruination. The Medusa could turn men to stone. Even worse than her mug was that of the basilisk, a large serpent-like creature whose mere glance could kill at a distance. Naturally, no one ever described it very well, since there were no survivors. As folks back then knew, human imagination has always been the best tool to create the most profoundly frightening images.

SEEING THEIR WAY CLEAR

Being without optical protection such as safety goggles and eyeglasses, long-ago workers throughout the Mediterranean basin had a natural dread of industrial accidents. Blacksmiths and other metal workers may have worn a patch over one eye to prevent blindness in both, should sparks or metal fragments hit them. Some scholars theorize that the plethora of Greek myths

Being a cyclops sucks. I've got no depth perception; where are those stinking lambs?

about one-eyed giants called Cyclops, who forged weapons and thunderbolts for the gods, correlated to actual practices and mishaps of human blacksmiths.

More realistically, the Cyclops stories may have gotten their start when the first ancient Greek or Italian ran across what seemed to be terrifying proof. In parts of Italy, on the Greek mainland, on islands like Samos and Crete, long-ago farmers occasionally turned up gigantic skulls; sometimes floods exposed them or earth tremors uncovered them.

Although twenty-first-century scientists now identify such skulls as extinct species of mastodons or mammoths, their long-ago discovers easily could have drawn the conclusion that they came from one-eyed giants. The hole where the trunk of this ancient ancestor of the elephant protruded looked ominously like the socket of a single huge eye. Look at an elephant skull and you'll be a believer. Over time, these finds could have reinforced Cyclopean myths, since they were viewed by generations of people. Ancient bones, oversized skulls, and other curiosa were frequently taken to the nearest temple and put on display.

The most skin-crawling relationship between real-life humans and mythical Cyclops, however, had to do with a powerfully toxic plant called white hellebore (*Veratrum album*, in the Liliaceare family). For centuries, tiny amounts were given to purge patients. Oh boy, talk about a stimulus package—this one worked at both ends. White hellebore actually became fashionable, used to treat a host of ailments from giddiness and palsy to epilepsy, tetanus, and white leprosy. As Pliny noted in his encyclopedia, scholars took it regularly to sharpen their brains. It was also prescribed to patients for mental problems. Besides its propensity to provoke vomiting in a stunning variety of colors, white hellebore contained the alkaloids cyclopamine and jervine. It's now recognized that both are teratogens, which may cause one-eyed Cyclopean birth defects.

Those who escaped the horror of one-eyedness due to accident, illness, or hellebore faced other ordeals. Even run-of-the-mill ocular disease or trauma demanded eye doctors and surgeons with sure hands, since painkillers and other mollycoddling measures were scarce. Dosages of opium and the like were hard to calibrate, so ham-handed anesthesia could add another soul to the fatalities column.

Spurred on by the sheer quantity of battle wounds, off-duty eye accidents, and grisly infections that soldiers were afflicted with while in the military, medicos developed precision instruments to work on the eye, along with techniques that still elicit admiration from optical experts of our day.

Cataract removal, made famous by the Roman surgeon Cornelius Celsus, required steady hands and a strong stomach on the part of both doctor and patient. It also called for more than a bit of luck, since negative outcomes, including blindness, were not uncommon.

A cataract does its dirty work by keeping light from reaching the retina at the back of the eyeball, eventually clouding the entire lens that rides behind the iris and pupil. Doctors in Roman times used a thin needle to move the whole structure, pushing it down and away from its original site. Although associated with Celsus, this procedure—called couching—was not invented by him or the Greeks but in all probability came from the technically advanced eye specialists of India.

Through archaeology, it's become apparent that Roman eye doctors made impressive advances regarding this exacting operation. In the 1980s, archaeologists working in France (formerly Roman Gaul) found a mysterious cache of instruments. Among the tools were delicate bronze instruments composed of fine retractable needles tightly encased in metal tubes. They were finally identified as a cataract removal kit.

To perform the surgery, a fine needle was directed into the cataract to break it up. Once the needle was withdrawn, the surgeon used the tube to suck out the bits, extracting the lens instead of simply moving it aside. What the patient must have gone through boggles the mind—his hands tied, his cranium in a headlock by the physician's assistant.

A full description of this innovative albeit excruciating surgery was eventually discovered. No wonder it took time to surface; the passage was found amid the nearly three million words written by that verbose genius of medical lore, Claudius Galenus. Galen had the stomach for harrowing medical procedures; before he made it onto the imperial physician circuit, he'd spent years patching up gladiators with ghastly wounds.

EARTHQUAKE STORMS, GIANT BONES

Geologically speaking, Mediterranean lands have been rocking and rolling for millennia. The cause: huge tectonic plates grinding against one another, provoking earthquakes with unpleasant frequency. The terrain bristles with volcanically active peaks such as Etna and Vesuvius, fed by magma deep within.

Greeks and Romans were painfully aware of the trembling earth they stood on, the land masses that mysteriously appeared and disappeared. They tried to placate the gods in charge of such things, but without much success. With grim regularity, they got hammered by geological disasters.

In 373 B.C., a huge quake, followed by a tsunami, destroyed several cities on the Gulf of Corinth. Seismically speaking, however, the first century A.D. was far more horrifying. During the reign of the Roman emperor Tiberius, twelve major cities in Asia Minor were wiped out in a single night. Shortly after that, a massive tremor shook part of Crete's coastline into the sea. During the same period, the famed encyclopedist Pliny the Elder described ten islands that had emerged abruptly within his lifetime. Pompeii and neighboring towns were still struggling to rebuild from a severe quake in 63, as well as smaller tremors for years afterward, when Vesuvius cut loose in a big way in August of A.D. 79. Mankind during that queasy century may have undergone what present-day scientists call an "earthquake storm," a multiquake run of seismic sequences.

Another quake storm lasting fifty years may have occurred in the late Bronze Age (1225 to 1175 B.C.), causing the destruction of Mycenae and Troy, among other urban centers. In August 1999, the same region experienced an earthquake storm that left seventeen thousand dead when a 7.4

quake hit. Three months later, a 7.1 magnitude tremor struck south of the first, both along the 650-mile Anatolian Fault.

This part of the world was also subject to slow-motion disasters. Earth movement, river silting, and rising seas made shorelines change. Certain seaside cities lost their waterfront status—sometimes by miles—and eventually were abandoned. The metropolis of Ephesus, for example, was once so close to the sea that waves lapped at the steps leading up to the wondrous temple of Artemis.

In Homeric times, earthquakes were blamed on the gods' anger, end of story. Later thinkers from Aristotle to Zeno all had a go at explaining the causes. Aristotle said quakes took place when winds broke through spongy or cavernous earth. The Stoics also endorsed the flatulence theory, blaming quakes on moving air. Hardly anyone speculated about fire within the earth itself. These early scientists wrestled to find the truth but failed to guess at the tectonic plates sliding beneath their feet, the millions of years of earthly history we now know about.

Earthquakes aside, legend-spinning common folks had their own explanations about the deep past. Drawing on folktales and Greek mythology, most believed that in previous ages, a larger brand of deities, along with a race of giants, had inhabited the earth. They agreed that certain semidivine mortals had once been supersized as well, especially heroic overachievers like Hercules and Pelops.

The existence of a race of superwomen called Amazons was one way in which the Greeks "explained" their deep past.

The Greeks (and their Roman borrowers) looked at the passage of time as they would a flowering plant, vigorous in youth but headed inexorably toward maturity and death. As time's clock ran down, the human race did, too, growing weaker and smaller than the giants who'd dominated the earth in its golden age. Therefore, when an earthquake dislodged the femur of an extinct giant giraffe, the partial skeleton of a Pliocene species, or the skull of a prehistoric beast, such tangible mementos confirmed legends, "proving" the existence of giants and semi-divine heroes.

Remember the quake that wiped out twelve cities in a single night? It did damage in Sicily and near the Black Sea as well. In places where the earth opened up, huge bodies appeared. The skeletons of heroes, many said. A delegation was sent to Emperor Tiberius with a monstrous tooth from one of the bodies. A cautious sort who didn't want to commit sacrilege, Tiberius had an in-house carpenter whip up a model showing the size of the head that would be needed to hold that tooth. When Tiberius got a glimpse of the mock-up, he hastily sent the tooth back home for proper burial.

Classical folklorist and author Adrienne Mayor has made a lifelong study of paleontology and its relationship to Greco-Roman folklore, linking myths to the finds of ancient bones, their interpretation, and their use. As she points out, around the seventh century B.C., people began to take their legendary heroes seriously. Cult worship sprang up. At these shrines and heroes' tombs, worshipers wanted to possess evidence of that hero, proof of his deeds. Such artifacts were looked on as lucky, as holy protection, as power.

To long-ago people, femurs and other bones from extinct animals like the mammoth often looked like giant versions of human bones.

About 560 B.C., when the Spartans asked how to defeat their enemies the Arcadians, the Delphic oracle told them to locate the bones of the hero Orestes—and they succeeded in doing both. Galvanized by such events, other city-states began their own quests. Athens went after the bones of its legendary founder Theseus; Kimon, the general who dug them up on the island of Skyros, became a hometown hero. City-states went batty over bones from the heroes of the Trojan War. In Asia Minor, locals were sure they'd found the grave of Ajax, the proof being his enormous kneecap, the size of a boy's throwing discus.

Fossil souvenirs of monsters from mythology also became sought after. Since bone finds often lacked skulls and were piecemeal, they might even be identified as warrior heroines such as the Amazons. This wasn't as crazy as it sounds. A femur from a mammoth looks much like a human legbone—just more mammoth. Today we view the skeletons of dinosaurs, prehistoric elephants, and other large extinct mammals tidily assembled in correct order. The Greeks and Romans saw a jumble of bones and sought to make sense of them by arranging them in humanoid order.

At times, this dramatic bony "proof" was reburied; more often, it was reverently displayed at shrines and tombs. This body of imaginative evidence tended to create a climate of belief in the reappearance of the giants who'd walked the earth—especially in tense times of seismic upheaval.

For instance, in the ominous, tremor-filled lead-up to the A.D. 79 eruption at Mt. Vesuvius, sincere locals in southern Italy, possibly hysterical with fear, started "seeing" things. As historian Cassius Dio said in his *Roman History*: "This was what befell. Numbers of huge men quite surpassing any human stature—such creatures, in fact, as the Giants are pictured to have been—appeared, now on the mountain, now in the surrounding country, and again in the cities, wandering over the earth day and night and also flitting through the air. After this, fearful droughts and sudden and violent earthquakes occurred, so that the whole plain around seethed and the summits leaped into the air."

Leaping lizards. Leaping summits. Whew.

MY BIRTHDAY'S VI-XIV; WHAT'S YOURS?

Throughout their early history, the Greeks and Romans used an additive system of numbering. The thrifty Greeks recycled certain letters of their alphabet into numbers: the letter I for 1, D (delta) for 10, H for 100, and X for 1,000. For 5, they used the pi symbol or, when they were feeling flamboyant, another symbol resembling the Greek G.

The Romans also used their own alphabet as a numbering system, which made life a bit easier for stonemasons. Anything permanent—from an imperial pronouncement to a tombstone—got chiseled in stone. In their numbering system, scribes and stone masons wrote I for 1, V for 5, X for 10, L for 50, C for 100, D for 500, and M for 1,000. Call it clunky, but it sufficed for nearly M years.

After the Roman Empire slowly crumbled, medieval scribes began using certain subtractive conventions. These sneaky shortcuts had been used off and on by scribal slackers for some centuries; now deregulation opened the door further. Henceforth, clerks, monks, and other scribal workers agreed that it would be perfectly kosher to write IV instead of IIII for 4, IX instead of VIIII for 9, and so forth. This kept hand cramps and carpal tunnel syndrome from running rampant in those unheated castles and monasteries.

You'd do the same when faced with a number like 949. In the old additive system, it took fourteen letters to write: DCCCCXXXXVIIII. Now it became a clean-cut CMXLIX. Today we continue to use Roman numerals in the medieval style, except when we don't want to, as on the faces of fancy big clocks like Big Ben.

When and where else do Roman numerals still find gainful employment? On analog clocks and wristwatches (remember those?); in the preface pages of a book; and as ordinal numbers, such as Pope Paul XXIII and Henry VIII. They're used to number film and television sequels, from *Star Trek VIII* to *Friday the 13th Part X*.

Roman numerals sneak in when you least suspect them. At the end of films and TV shows, ever notice when the copyright date in diamond type crawls by? By using MCMIVII instead of 1942, they stand a fair chance of confusing viewers. The most popular use continues to be sequences of events, from the Olympic Games to Superbowl XLV—a number almost big enough to give one a nice warm sense of history.

To us, the glaring omission in the whole ancient numbers game was zero. We're so used to it; how could the Greeks and Romans do math without it? As it transpires, the notion of zero (and what's called positional notation) began many centuries before the Romans or Greeks. Dreamed up by the inventory-obsessed Sumerians of Mesopotamia, the zero may have been picked up as a curiosity by Alexander the Great and taken to India.

Among Greeks, the zero had a dubious reputation for centuries. They linked it to alchemy and the ouroboros swallowing its own tail.

The Indian savants welcomed the zero with joy. Not so the cultures of the West. Pythagoras, as respected a philosopher as one could find, gave the zero a thumbs-up, calling it the perfect form, the originator and container of all. His enthusiasm utterly failed to persuade. Zero or nothingness became associated with the ouroboros, the ancient symbol of the snake swallowing its own tail. As a result, only ancient alchemists and other dubious characters fooled around with zero.

Well into medieval times, Western cultures were still zerophobic about this symbol, surrounded by myth and mystique. Today the zero is out of the bag with a vengeance, being one of the two components of our ubiquitous binary code.

Speaking of myths: moderns still cling to a mathematical myth that refuses to die. Its discoverer, the Greek mathematician Euclid, dryly called it "the extreme and mean ratio." Others referred it by the softer label of golden ratio or GR. As an equation of measurement, the GR ratio of 1.6183399 (plus infinitely more digits, like the equally messy pi) has always interested mathematicians because it pops up in pentagrams, the Fibonacci sequence, and other places. It's also found in some forms in the natural world.

In the fifteenth century, an Italian mathematician swooned over GR, renaming it the "divine proportion." That ignited the hype, which climbed steadily until 1835, when another sensitive math nerd worshipfully dubbed the ratio the "Golden Mean."

Cool labels notwithstanding, the rapturous praise GR gets for having mystical properties of "perfect proportion" is humbug. Back in the first century B.C., writer-architect Vitruvius did indeed unskillfully compare parts of the human body to temple building, but the dimensions of his hypothetical human would never be found in his time or ours. Temporal association with Vitruvius is why famous structures such as the Parthenon and the pyramids are lazily cited as "perfect" examples of the GR.

As if this weren't bad enough, Leonardo da Vinci had to get embroiled. Leonardo, who was deeply interested in proportion, read old Vitruvius and got the idea of doing a drawing of a naked male within a circle and square, which he called "Vitruvian man." He didn't say that the human form displays the golden ratio or that his drawing had nailed it. In fact, Leonardo said nothing; other people said it for him.

Repeat a story long enough, print it untold times as a factoid, send it around the Internet a few trillion times, and it sets as firmly as Roman concrete. The hapless golden ratio equation is now doomed to an inaccurate immortality.

SCAPEGOATING THE POLLUTERS

Two thousand years ago, pollution was a religious offense, a crime against the gods. The Greeks pegged it first, calling it *miasma*, which meant something akin to "bloodstain."

At times, it could be a careless act that physically defiled or degraded something—pollution in our modern sense of the word. For example, Emperor

"Give a hoot, don't pollute," followed by a sacrifice, was the ancient way of dealing with the crime of pollution.

Nero once bathed naked in his city's aqueducts, the precious source of Roman drinking water. Why? Because he could. His act drew great condemnation for being both impious and filthy.

A laundry list of taboos existed to prevent miasma. It was considered pollution to cut down any tree in the sacred grove around a temple. Few people dared to flout these taboos—although the Roman senator Cato did. Despite his much-praised reputation for ethical behavior, he once advised some landowners who wanted to get around the tree-cutting moratorium: "Make a prayer to the 'to whom it may concern' deity—then whack it."

Humans were warned of possible miasma violations in other ways, such as omens at religious sacrifices. After an entrail reading in the first century B.C., a Roman dictator named Sulla was warned that his wife's illness would pollute his household. Instead of stepping down, he chose to divorce his allegedly polluted wife. She may have gotten off easy; being married to Sulla could not have been a stroll in the park.

Polluters in this early sense of the word usually faced a severe penalty: becoming a scapegoat. Like miasma, that word had real horsepower (or would that be goatpower?) back then. To be scapegoated had far worse implications than getting fired or bombarded with harsh criticism. It could mean being driven out of one's home city, becoming an exile. In some cases, scapegoat status meant a period of house arrest, followed by a ritualistic death carried out by religious leaders and witnessed by the public.

Many times, the scapegoat was not really the perp but an individual upon whom the dire implications of the polluting act could be dumped. In ancient Italy and Greece, this could mean a marginal person: a criminal, a slave, a person with deformities or birth defects. Or even a temporarily polluted individual, such as a woman who'd just given birth.

Some anti-miasma measures were performed annually. For example, in Chaeronea, the Greek hometown of the writer Plutarch, the city magistrate beat a slave with branches from a certain tree, then pushed him out of doors with the words "Out with hunger, in with wealth and health." In each household, locals followed suit with a similar ceremony.

Romans followed a similar scapegoating tradition every March 14, when they chose an old man, dressed him in animal skins, and drove him out of the city while beating him with long white rods.

These relatively humane examples were outnumbered by harsher rituals in other times and places. Whenever an epidemic or plague hit the Greek colony of Massilia (present-day Marseilles in southern France), the populace called an assembly together, asking for a male volunteer. It could be anyone; the most likely volunteers were malnourished men. Once the man was chosen, he'd be maintained at the city's expense for a full year, wining and dining on nothing but the best. When the year was up, they dressed him in foliage and sacred garments. The citizens, reciting prayers that all their evils might fall upon his head, led him ceremoniously through the city. When they got outside the city gates, the scapegoat was stoned to death. By this time, the plague that had started the whole process might have burned itself out, thus "confirming" the efficacy of the scapegoat technique.

When it came to literal pollution, the ancient world had its troubles as well. As top toxic sites, mines and their by-products from extracting and smelting copper, tin, sulfur, mercury, iron, gold, silver, and lead headed the list.

Silver was in heaviest demand—especially for coinage. Its ore was often found with lead, another heavily used commodity. To extract 1 ton of silver, 100,000 tons of rock had to come out of the earth. Mining contaminated streams and aquifers, while slag from the operations polluted the ground. Nearby communities got exposed to toxic fumes and downstream waste, which also degraded their lands under cultivation and the wild ecosystems they depended on. The extractive industries took a huge human toll. Some mines were even vaster than today's operations; the silver mines near Athens had 87 miles of tunnels. In Roman Spain, in just one mine, forty thousand workers toiled in the year 179 B.C.

In urban areas, air and noise pollution impacted daily life. City smog enveloped Rome; locals called it "infamous air." Besides wood and charcoal smoke, the air reeked of dung and urine, used extensively to tan hides, to clean wool, and in other processes. Aqueducts and wastewater management helped, but a percentage of the population did not have access to sufficient clean water or adequate sewage disposal.

Although they used magical thinking (and cruel scapegoating) to combat miasma, the people of Greco-Roman times were groping toward the truth. Then as now, polluting the earth was an offense against the gods, against humanity's children.

FROM BLOOM TO BLADE

Going for the gold at the ancient Olympics was all very well, but when it came to warfare, the Romans went for the iron. As encyclopedist Pliny was hawkishly fond of saying, "For slaughter, it is even more prized than gold."

Early on, alert citizens noticed that iron in the form of meteorites occasionally dropped out of the sky. In 53 B.C., for example, the area around the boot of Italy received a showy meteor storm that deposited a number of meteorites resembling iron sponges. Still, the sporadic nature of meteor showers made it impossible to forecast the number of swords (or plowshares) anyone could produce in a given year.

Iron had proven useful for a growing list of items: nails and clamps to hold marble columns together; iron bars to strengthen walls, vaulted buildings, concrete piers; keys, locks; work tools of all sorts; hobnails for Roman soldiers' sandals and armor for their bodies; swords and knives; iron rings worn by freedmen and freedwomen to advertise their nonslave status; fancy iron brooches, or fibulae, favored by women, especially since no one had thought to invent buttonholes or Velcro yet.

From earlier cultures in Etruria, Greece, and the Caucasus, Romans learned that reddish iron ore could be found almost anywhere but the top-notch stuff

Naked blacksmiths? Artistic license, perhaps. They usually wore leather aprons to protect themselves from sparks and intense heat.

for quality swordmaking was scarce. Always methodical, the Romans assessed the situation. Since the lands they already controlled, including the islands of Elba and Sardinia, lacked enough good ore, it became imperative to fight wars over iron in order to have iron for more weapons.

With that in mind, around 200 B.C. Romans went after the incredible mineral wealth of the Iberian peninsula. Before you could say *veni, vidi, vici,* Spain found itself conquered in 197 B.C., with a fair number of its sturdier citizens repurposed as subterranean excavation specialists. Mining slaves, in other words.

Other finds followed in Gaul and Britain. One especially rich discovery that became a political takeover occurred in early imperial times: the area of Noricum (present-day Austria and Bavaria), whose Alps were loaded with high-quality ore.

A finicky substance, iron required more attention than other metals. Romans solved the biggest mystery when they learned there was no need to get iron to its melting point of 1,535 degrees Fahrenheit in order to work it. Just as well, since reaching, much less maintaining, that temperature was quite difficult, given their technology.

Being able to work iron by direct reduction brought about the advent of the small, pyramid-shaped bloomery furnace. As Roman blacksmiths found, the real trick was to select ore that wasn't too dense and had the right ratio of silica glass to iron. Maintaining the proper temperature, keeping an even heat, and getting the right amount of oxygenated air by using bellows were also key.

As iron reached 800 degrees, it magically turned red and spongy, a quality called bloom. Inside the glowing mass, the iron particles were bonding loosely. Tiny channels began to open inside it, filled with silica glass and other impurities that had been imprisoned in the ore.

When the bloom reached its moment of truth, the blacksmith removed the piece from the furnace and begin to hammer it, reshaping the bloom, driving out the glass slag, then quenching the piece in water. He might reheat and repeat, tempering the piece as he shaped it. He could continue this final stage until the metal cooled. The sizzling sound of quenching the iron was familiar to blacksmiths even in Homeric times. *The Odyssey* itself contained a grisly scene where the heroes put out the eye of the Cyclops: "Just as when a smith plunges into cold water a mighty axe-head to temper it—for this is what gives strength to the iron—and it hisses violently, just so did his eye sizzle around the olive branch."

To produce a high-end piece such as the two-edged Roman gladius, the signature weapon of the legions, took special care. A sword could be forged from a single bloom, but more often the blade was a series of welded strips of iron that the blacksmith struck while hot until they formed a unit. By reworking portions of the iron, he could make parts of the sword, especially the cutting edges, have a higher carbon content and thus be steelier and more durable than other parts.

Archaeologists in our era have carried out long-term projects in experimental archaeology, building bloomery furnaces and attempting to re-create the conditions in which ancient Roman blacksmiths worked. Among other things, they've discovered that charcoal, the ancient fuel of choice used to fire such furnaces, let the iron ore absorb more carbon. Much of the accurate and detailed information we have today about ancient iron and swords of the era comes from their deep research.

Although Romans approached ironmaking with commonsense empirical trials to learn what worked and what didn't, the same could not be said of iron's usage in the larger culture—and the wonderfully strange beliefs held about the metal. Rust, for example, was considered a curative for a great many things: scabbiness, pimples, hangnails, gout, and creeping ulcers. Add vinegar, and you could take care of your anal swellings as well. Scale of iron worked to stop hemorrhage. Make a wet plaster with iron, add bee pollen, pounded copper, wax, and oil, and you could even put new flesh on bones that had been denuded. Or so it was claimed.

To protect infants against noxious drugs, doctors urged parents to draw a circle around the child with a piece of iron. Problems with nightmares? Just extract a number of iron nails from tombs, make a fence of them in front of your threshold, and presto—sleep like a baby. These magical medical applications make one long for more information. How did they pound iron nails into marble tombs in the first place? Were they a decorative element? Or a magical barrier to keep family corpses free of anal swellings? The minor mysteries of history, sometimes as compelling as the big moments.

AN ALL-AROUND, ALSO-RAN ARCHITECT

The Greeks had harsh ways of dealing with architectural cost overruns. Author Marcus Vitruvius Pollio, our only extant source for architectural nuts and bolts, applauded a law said to be on the books in the Greek city of Ephesus. "When an architect [of that city] accepts the charge of a public work, he has to promise what the cost will be. His estimate is handed to the magistrate, his property is pledged as security until the work is done. When finished, if the outlay agrees with his statement, he is complimented with decrees and marks of honor. If no more than one-fourth has to be added to his estimate, it is furnished by the treasury and no penalty is inflicted. But when more than 25 percent has to be spent in addition, the money required to finish it is taken from the architect's property."

This excerpt was followed by Vitruvius' own wistful comment: "Would to God that this were a law also of the Roman people, not merely for public but also for private buildings."

Clearly a man who'd fumed more than once over slippery colleagues and out-of-control project expenses. Vitruvius was a plainspoken fellow from Verona, it's thought. He spent years as a military architect and engineer under Julius Caesar, campaigning from Gaul to North Africa. After Caesar's shocking assassination, he transferred his loyalty to Caesar's grand-nephew (and adopted son) Octavian.

Once Octavian became the Big Mozzarella of Rome in 27 B.C., Vitruvius began writing his ten-book opus. A good storyteller, he was careful to credit his sources. In *De architectura*, he praised his parents for giving him an excellent education, citing the value of history, art, music, philosophy, biology, geometry, math, and astronomy to his chosen profession.

Philosophy, Vitruvius claimed, not only made architects more courteous, just, and honest but also taught them physics. He also endorsed environmental sanitation: "The architect should have a knowledge of medicine on account of the questions of climates, air, the healthiness and unhealthiness of sites, and the use of different waters."

Architects of his day wore far more hats than today's specialists. Vitruvius had to juggle the tasks of city planner, landscape architect, artist, engineer, and contractor as well. As he noted, "There are three departments of architecture: the art of building, the making of time-pieces, and the construction of machinery." Timepieces? Yes indeed—he elaborated on sundials and water clocks, tying in related topics from the weather to the zodiac.

One of the sections most prized by modern researchers is on machinery—much of it data found nowhere else. (Centuries later, Vitruvius' descriptions of the distance-counting odometer, the Archimedes helical screw, and other inventions were studied and prototyped by Leonardo da Vinci.)

In his career, Vitruvius went from army architect to civilian life, but commercial success did not follow. His architectural projects were few. This book, however, was sure to bring him business, he thought. First of all, he'd dedicated it to Rome's new emperor. Second, after years of civil war and civic neglect, the emperor planned to rebuild Rome, promising a shining city of marble to replace a dirty city of bricks. Spectacular architecture to please the Romans—that was the ticket. And his visionary book, a how-to about classic Greek architecture, would lead the way. The emperor couldn't help but love his three governing principles, *firmitas, utilitas, venustas*—strong, useful, and beautiful.

It was a splendid plan. Only the gods knew why it failed to work. Emperor Octavian Augustus indeed launched an enthusiastic building campaign. Vitruvius, however, got bupkes. No commissions, no glamour jobs. Not even one lousy contract in Rome to build a temple, a public bath, or even a stinking set of latrines.

Maybe the emperor thought he was too old. Granted, he was in his seventies, unhandsome, short, and not as strong as he once was. Was it a crime that he'd tried for a little sympathy by mentioning those facts in his book? Maybe the emperor thought he should retire, content to live on the nice little veteran's pension he received.

He still cherished hopes that future generations might appreciate him. In his book's preface, he poured on the pathos, saying, "I've achieved little celebrity. Once these volumes have been issued, however, I hope that I will be renowned to future generations."

Marcus Vitruvius got his wish. His writings had the luck to survive the vicissitudes of time and did just what he'd hoped. It did take a while—fourteen centuries, to be exact.

During the Italian Renaissance, celebrated architects Leone Alberti and Andrea Palladio ate up Vitruvius' words and secrets, calling him "master and guide." From colonial times into the modern era, his books have remained an invaluable tool for architects, historians, and archaeologists to help reconstruct the classical ideal. *De architectura*, his labor of love, did not possess the most golden of prose, but it was studded with gems that have grown ever more valuable with time.

UNDER THE VOLCANOES

It's always Vesuvius, Vesuvius, Vesuvius. What about Stromboli? What about Mt. Etna, for Hephaistos' sake? Italy bubbles with impetuous volcanoes.

Not just Italy, either. The topography around the Mediterranean boasts many active peaks, sitting at the continually colliding edge of two tectonic plates, the African and the Eurasian. As a result, volcanic magma erupts periodically, as do disastrous earthquakes and tsunamis—such as those that destroyed the Minoan culture on Crete and did a number on the Greek island of Thera.

The geology of this region is an open-air lab of titanic forces. Just as intriguing, however, is the long history that humans and volcanoes share. What brought such population density to these areas? Volcanic soil. Rich soil, where fabulous grapes and a multiplicity of crops could grow like mad in places such as southern Italy and the island of Sicily.

Another volcanic gift, gratefully exploited, was a reddish gray dust called pozzolanic ash. Found in the vicinity of Vesuvius and Etna, it bound together

lime and aggregate to make an astounding concrete that set rapidly—and did it just as well underwater. Its discovery in the second century B.C. turned Rome (and other cities) into a building powerhouse, a contractor's dream. Didn't hurt the ambitious building plans of future emperors, either. Not only could jetties and arcaded aqueducts be constructed, but so could cheap high-rise apartments. Concrete nearly solved Rome's low-income building shortage—and that of other places, too, since pozzolanic ash could travel by ship anywhere with ease.

The eruptions of Mt. Etna weren't always bad luck, either. In 396 B.C. it exploded, its eruption fortuitously coming to the rescue of local Syracusans by scaring off an attacking army from Carthage. Most of the time, the chocolate brown shoulders of the great volcano barely shook; it hummed away, dribbling rather than spewing lava. The volcano's relatively gentle nature even made it a climbing destination. Vigorous travelers struggled up its

10,900-foot slopes to gape into the craters. One emperor even hiked it; Emperor Hadrian summitted on February 5 of the year A.D. 62. Etna also had a weirder, metaphysical appeal: its caldera became the site of several leap-in suicides, the most famous being that of physicist-philosopher Empedocles around 432 B.C.

A frightful Greek legend surrounded Etna. Under its simmering flanks, the wind god Typhon lay buried. Its monstrous form had a human head, huge viper coils, and vast hands tipped with hundreds of dragons. An enemy of the Olympian gods, he and Zeus battled for supremacy—and Typhon lost. Thus the smoke signals and volcanism of Etna were first explained in poetic and mythical form.

Stromboli, a hyperactive volcano doubling as an island, was one of the Aeolian Islands off Italy's toe. In ancient times, ships could see its eruptions for miles. It served as a beacon for mariners, earning the nickname "lighthouse of the Mediterranean."

Vesuvius itself was quiescent for long intervals. One period of calm occurred in 73–72 B.C., which coincided with an era of high alarm among human populations throughout Italy. The epicenter of their fear? A man called Spartacus, a Thracian POW sold into slavery as gladiator fodder. He was in training when he and seventy-seven others broke out of gladiatorial school in Capua, near Vesuvius. Stumbling on a source of weapons, they quickly became a force to be reckoned with. Although modern accounts often call him the sole leader, Spartacus was one of three men elected by the group. Their ranks swelled by runaway slaves, thrill seekers, and ruffians, the band fought its way through urban areas to Mt. Vesuvius, where they made a camp on the flat part of its summit. (That level campsite and much else would disappear 152 years later in the eruption of A.D. 79.)

Finally galvanized by the Spartacus phenomenon, the Roman Senate took action. Already stretched thin by wars in Spain and Asia Minor, they sent three thousand raw troops with an equally green general named Glaber after the gladiator rebels.

Glaber was sharp enough to spot Spartacus and company up on the mountain, and figured he had them surrounded. At that time, however, Vesuvius had vegetation growing thickly up its flanks. During the night, the rebels wove ladders from vines, descended the mountain on the opposite side, and got behind the Roman lines. Glaber's inexperienced troops panicked and ran for it. The forces of Spartacus looted their camp and grabbed their abandoned weapons, leaving their Vesuvius refuge to march forth.

They ravaged chunks of Italy and beat a succession of Roman generals in an extended but doomed campaign that ended with Spartacus' death in battle. His band was extinguished in various unpleasant ways; thousands of them were crucified along the Appian Way to serve as a grim warning to other slaves.

Although tremors hit Rome and cities in Asia Minor during this period, Vesuvius continued to show a placid face until February of A.D. 63, when a bad earthquake in Pompeii and Herculaneum did significant damage. To all appearances, Mt. Vesuvius had not misbehaved. Nevertheless, by May of A.D. 64, more bad omens appeared—one of them named Emperor Nero.

That May at the big theater in nearby Naples, the emperor arrived for his first public performance as a singer. Oblivious to his third-rate voice, he launched into his repertoire, singing the hours-long tragedy *Niobe*. In the middle of his gig, a medium-sized earthquake struck the theater, its epicenter the nearby island of Ischia. Nero kept on wailing. The audience, equally terrified to leave or remain, kept their seats. Once he finished his set, the relieved crowd swarmed into the street. Behind them, in an eerie curtain call, the theater collapsed into dust.

Most people were horrified by that ominous incident. Nero, on the other hand, probably interpreted it as thunderous applause from the gods for his artistry. He and his Pompeiian wife, Poppaea, soon went on to Pompeii to be welcomed there. Neither Poppaea nor Nero would be alive when the final curtain came down for Pompeii and Herculaneum, obliterating the extravagant, frenetic life around the Bay of Naples on a sultry August day in A.D. 79.

PASSIONATE POLYMATH

Two and half millennia ago, knowledge seekers rarely confined themselves to one field of endeavor, considering the ideal intellectual pursuit to be that of wide-ranging generalist.

Empedocles was such a person. A native of Agrigentum, Sicily, he lived in a rich Greek city of two hundred thousand souls, whose landscape was ridged with huge honey-colored Doric temples. A true polymath—poet, playwright, orator, and doctor who wrote a lengthy book called *Discourse on Medicine*—Empedocles came from a distinguished and wealthy background but had a restless mind. He soaked up learning, mastering philosophy and becoming a noted teacher. Galen called him the founder of the Sicilian medical school. Locals hailed him as a miracle worker. We might call him a pioneer in evolutionary biology and ecology.

Sicily, home of philosopher whiz Empedocles, was rich in rivers—and reverenced river gods like this one.

He thought of the universe around him as a poetic, endless recycling. As he put it, "There is no birth in mortal things, and no end in ruinous death. There is only mingling and interchange of parts, and it is this we call 'nature.' "

Wandering through the forested wilderness that was Sicily of his day, exploring its rocky cliffs, its sandy coasts, he looked intently at the living things around him. His mind took a huge intuitive leap, guessing that many different species had once existed here—and some had been unable to survive the harsh conditions of their surroundings, while more adaptable creatures took their places. Among the first to speculate about natural selection and survival of the fittest, Empedocles also noticed that many species of flora and fauna seemed to prefer niche environments, and he wondered why.

Unlike other rich-boy deep thinkers we could name, Empedocles had a pragmatic bent. His home island, rich in rivers, also had swamps alive with insects, including vast numbers of mosquitoes. More than a few carried malaria, presenting a serious health problem for the islanders. Sensing the jeopardy of the townsfolk of nearby Selinus, Empedocles used his own wealth plus the help of competent hydraulic engineers to divert the course of two rivers to the area. Needless to say, he became a hero to locals, whose rates of child mortality, malaise, and mysterious fevers dropped appreciably. Did he realize the cause was insects rather than something in the air? It's possible. Three centuries after Empedocles, a respected Roman historian named Varro claimed in his books that small flying beasties, not "bad air," caused malaria. (The word *malaria* is Italian for "bad air.") Varro had to have picked up this information from an earlier source, since he did not carry out any research himself.

In his scientific inquiries, Empedocles naturally stacked up some gigantic misses. For instance, he dismissed the sun altogether and then had to grope for a way to explain daylight. He did so with an unconvincing rap about bright and dark hemispheres circling the earth. Nobody bought it; still, Empedocles lived in a fluid time, when natural philosophers could imagine six impossible things before breakfast and no one shouted them down.

Although described as gravely dignified, he espoused a lovely, lively doctrine. He believed that six roots or elements made up the world—fire, water, earth, and air, plus a force he called friendship that united the elements, and a force called strife that repulsed them, a bit like the positive and negative poles of a magnet.

From his close studies of Orphic and Pythagorean philosophy, he came to have a firm belief in reincarnation and the transmigration of souls. He was quoted as saying, "And their continuous change never ceases, as if this ordering of things were eternal. The soul, again, assumes all the various forms of animals and plants. Before now I was born a boy and a maid, a bush and a bird, and a dumb fish leaping out of the sea."

All this erudition, and the man was a snappy dresser, too. He favored purple robes cinched with a spectacular belt of gold. But the sartorial items that really caught the eye were his bronze sandals or slippers. (Not solid bronze, we're guessing; bronze-colored or fancy metallic soles, perhaps.)

Empedocles developed quite a following, especially after the swamp-draining episode and his generosity in giving dowries to local girls from poor families. Then he won mass adoration after he saved the life of a Sicilian woman who had been in a trance for thirty days and was given up on by other doctors.

Right about now Empedocles might have begun to buy into his miracle worker reputation a little too much, or perhaps he thought it was time to field-test his reincarnation theory. Although no bloggers or television reporters were there to corroborate the story, it was said that he climbed Mt. Etna and jumped into its fiery volcanic heart in lieu of a tamer suicide. One sandal of his trademark bronze footwear was later found at the caldera's lip.

Like other much-admired figures who lived in the misty long ago, the life of Empedocles swirled with contradictions. By one account, he never left Sicily. According to another, he traveled to Greece to attend the Olympics and recite his book called *Purifications* to the crowd, afterward falling from a carriage and breaking his thigh. (How does one break a thigh, anyway?) There being no hip replacement specialists on hand, he died and was buried in Megara. Yet another report asserted that he just plain slipped into the sea and was drowned.

Frankly, the volcano jump sounds much more in the passionate, inquiring spirit of the man.

EARLY GOOGLING

Back in the third century B.C., a lot of important stuff had not yet been invented. Weekends, for instance. No one had bothered to come up with a way to describe big numbers, either.

A young Greek geek named Archimedes decided to tackle some of these much-needed items. The son of the astronomer Phidias, who probably introduced him to the love of science and numbers, he and his family lived on Sicily in the beguiling Greek city-state of Syracuse during its golden age.

The boy was a true scientist from the word *eureka*. Hated to get dressed, sometimes forgot to eat or wash up. According to an anecdote from a later historian, when young Archimedes started smelling too ripe, the servants toted him to the public baths. Instead of letting the bath attendant clean him with olive oil, he would continue working on geometry problems, using the oil on his body as a "blackboard" of sorts. His idea of fun was to exchange jokey letters with fellow eggheads, daring them to find proofs for the weird problems and geometric theorems he supplied.

Thirsting for more challenges, as a young man Archimedes cajoled his parents into sending him to Alexandria, 900 miles from Syracuse, to work with other math whizzes. That way, he got to mingle with fellow nerds at the Great Library and Museum, Alexandria's grand institution of research and higher learning, and study with the successors of Euclid, the geometry genius.

Although he might have preferred the pure science of abstract numbers, Archimedes soon got put to work by Syracuse's ruler Hieron II. Being a relative of the top man had its pros and cons; Archimedes was now obliged to turn his big brain toward practical problems. There was a certain amount of urgency from Hieron's point of view, since the First Punic War between Rome and Carthage was about to explode, with Syracuse caught in the middle. (He didn't know the conflict would last two decades, either, which was just as well.)

What with all the commuting between Alexandria and Syracuse, and the length of the war, and the military hardware and software he had to dream up, prototype, and test, Archimedes (while still sharp as a tack) found himself well advanced in age before he got to concentrate on the mathematics problems that fascinated him. Although he had yet to grapple with the thorny problem of weekends, he tackled his other goal: to calculate the number of grains of sand needed to fill the entire universe.

There were a number of obstacles in Archimedes' way. First he had to determine the size of the cosmos, as he knew it. Although other astronomers—including his dad, Phidias—favored an earth-centered universe, Archimedes leaned toward the notion of the sun at the center of the planetary system. An astute thinker named Aristarchus on the island of Samos, another epicenter of Greek geekism, had postulated just such a thing.

Second, Archimedes had to set up a variety of assumptions, the first being the size of your average grain of sand. He decided that it would take ten thousand grains to fit inside a poppy seed.

The worst chore wasn't visualizing the problem—it was what to call the numbers. Although Greek scientists and thinkers loved to talk about complex systems and enormous quantities, they had been terribly lax when it came to the notation system they used to talk about them.

There were no Arabic numerals in use, no numbers per se—just a tiresome method of borrowing certain letters of the Greek alphabet to stand in for figures. The Greeks did have a word for 10,000: *myriad*. The only other whopper-sized number that could readily be expressed was myriad myriads, meaning 100 million. (Myriad myriads is also the largest number named in the Bible.)

Archimedes set out to rectify this slovenly state of affairs. He did so by creating a numbering system of multiple sets that still works for mathematicians today. It allows the expression of figures up to and including a 1 followed by eighty quadrillion zeroes! A number with some gravitas to it—one that put the good old googol (a 1 followed by a mere one hundred zeroes) in the shade.

Moreover, he did this without any help whatsoever from that modest little circle we call zero. Intellectual mission accomplished, Archimedes gave his eight-page treatise the catchy title of *The Sand Reckoner* and went on to other mind-blowing quests.

Although the word sounds like it derives from ancient Greek geek-speak, it should be noted that the word *googol* didn't come into existence until 1940 or so, when the smartypants nine-year-old nephew of U.S. mathematician Edward Kasner coined it. Five decades later, a pair of boy geniuses at Stanford University reworked the spelling to google and the rest, as Archimedes would affirm, is history.

ANCHORS AWEIGH. OR MAYBE NOT.

So you think the Roman Colosseum was colossal? You should have seen the superships that used to prowl the Mediterranean and Black Seas.

Supersizing really got its start in Hellenistic times, after Alexander the Great died and his successor generals, including a standout named Ptolemy I, began to arm-wrestle over kingdoms and possessions, showing off in that boyish way of "mine's bigger than yours—and longer, too."

For centuries, Greeks had patrolled their waters and fought naval battles with 121-foot triremes, the sturdy, speedy (up to 7.5 knots for long periods of time) workhorses of the sea. Powered by rowers, they carried a large square sail (the secondary source of power) and three banks of oars. Now and then, some hotshot would build a four-banger or even a quinquereme, but that was as far as it went.

A hyperachiever named Demetrius of Macedon changed all that. As the Greek world looked on in appalled fascination, the man commissioned a whole series of grotesque vessels. An elevener. A thirteener. Neptune help us, a sixteener. By the time Demetrius was finally overthrown in 285 B.C., his rivals had caught the fever. Some built teeners; a few launched a couple of twenty- and thirty-banked vessels. Egypt's young Ptolemy II put together a fleet of 336 vessels, from triremes to thirtiers.

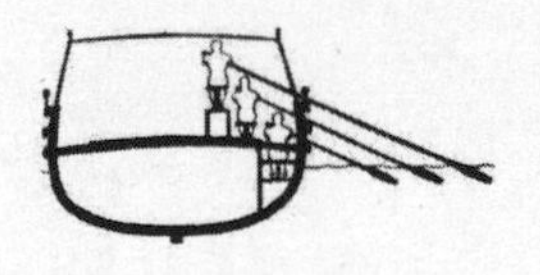

Before superships, Greeks relied on the trireme, a hardworking vessel powered by three banks of rowers.

Several generations of Ptolemies later, however, the race for supreme naval superlativity was won by a flabby hedonist called Ptolemy IV Philopator, who depleted the rich holdings of Egypt, Judea, Syria, North African Cyrene, and Cyprus he'd inherited to build a series of still talked-about superships. To warm up creatively, Ptolemy threw together a modest riverboat weekender, a floating pleasure palace of 13,500 square feet.

Then he got serious, commissioning the world's first (and only) fortier, a supergalley vessel that measured 57 feet wide and 420 feet long. Towering higher than a seven-story building, the elephantine ship was so vast it took four thousand men to row it—and another four hundred to man the sails and rigging. It was a two-hulled craft, in appearance a cross between a giant catamaran and an aircraft carrier. With oars.

After a mammoth struggle to get the ship out of its wooden cradle, it became clear that Ptolemy's nautical baby was not a whale but a dinosaur. It never saw service after its maiden voyage. One bright spot in the day's disaster: a bright Phoenician lad on hand at the time, seeing the crying need for a easier way to launch and store such vessels, invented the drydock.

Ptolemy had a salt-water rival named Hieron II, who tyrannized over Sicily's most brilliant city, Syracuse. Since his holdings were rich with wheat farms, he needed superships to haul grain. His first fleet was big but not mutantly so. Then Hieron heard about the supership fortier and lost his head. He turned to his staff of architects and engineers, which included his kinsman Archimedes.

Hieron began to harp at Archimedes: "You're supposed to be a genius. Why not build me a ship that combines the best features of a freighter, a royal yacht, and a warship?" Why not, indeed—the answer being that, like countless printer-scanner-faxer-copiers hitting landfills today, all too often multitasking machinery is not very good at any of its tasks.

With a sigh, Archimedes oversaw the construction of the good ship *Lady of Syracuse*. It was undeniably a beauty, a three-masted ship with twenty banks of oars. Archimedes made sure it had beautiful workmanship and the gadget-filled works: luxurious quarters for the head tyrant, twenty horse stalls, a 20,000-gallon water tank for humans, seawater tanks for live fish, a library, walkways with flowerbeds and grapevines, a hold that could carry 1,800 tons of wheat, salt fish, and wine, and little touches such as bronze tubs, a gym, and a shrine to Venus made of gems, ivory, and cedar.

Archimedes busied himself, arming the vessel to the teeth with cranes for missile hurling, catapults for stone throwing, and snazzy armored turrets for the archers. It was an exercise in futility. Like Ptolemy's ill-fated vessel, Hieron had overlooked one detail: the *Lady of Syracuse* was too big to dock anywhere except at Ptolemy's landing in Alexandria! Thus its first and last voyage was straight to the dock of his rival, a giftie from one marine megalomaniac to another.

As Rome became a power, more practical superships were built. Some were 180-foot grain ships that carried 1,000 tons of wheat and nine hundred passengers; others were vessels that lugged 200-ton marble columns and monster obelisks that a certain ruler just had to have for his redecorating schemes.

To show up the supership maniacs who'd preceded him, Roman emperor Caligula built two Costco-sized pleasure barges.

Fabulosity, however, still reigned among deranged CEOs of long ago. Two of the most marvelous floating behemoths ever built didn't touch salt water at all. They sprang from the manic brain of Emperor Caligula during his four years of rule and were later enjoyed by Emperors Claudius and Nero.

To give more privacy and scope to his creative perversity, Caligula commissioned twin houseboats for Lake Nemi. This lake south of Rome, nicknamed "Mirror of Diana," was surrounded by the wooded haunts of the goddess of the hunt and the moon. Around its tranquil waters, the houseboats would be rowed at a sensual crawl.

The larger of the golden pleasure barges had five keels and two decks of carved oak, stretching 235 feet long and 65 feet wide. On board were elaborate water delivery and bilge pump systems. Ball bearings—the world's earliest prototypes—drove a rotating platform. The ship came fully loaded with marble fittings, bronze art, mosaic pavements, mature fruit trees, and a temple. It was so weighty that later generations could not raise the vessel to the surface.

Instead, in 1932 the level of the lake was lowered 60 feet to reveal the Nemi ships and haul them ashore. Mussolini and a few favored Italians had a decade or so to enjoy them before the originals were destroyed in World War II. Today, with the support of the Association Dianae Lacus, shipwrights and archaeologists are busy creating a faithful replica of one of the fabled Nemi ships.

LEVERAGED OUT

Everyone recognizes the name Archimedes; something about a lever, a fulcrum, and that throwaway line, "Give me a place to stand and I will move the earth!" It's always sounded like clever but inexplicable hooey; what's missing is context.

As a youngster, Archimedes was considered the brainiac of Syracuse, itself an urban beehive of intellectual ferment on Sicily. As a middle-aged adult, however, yearning for nothing more than a nice little grant to work on his favorite math problems in peace, Archimedes kept getting interrupted by requests from his kinsman King Hieron II. Being the city's top tyrant, Hieron's "requests" were more pressing than your average pretty-please.

By this time, Hieron had gotten deeply into shipbuilding. Not just any ships, but supersized beauties. He and Egypt's ruler Ptolemy Philopator had a friendly rivalry as to who could produce the vastest seagoing monster of the deep. Naturally he embroiled Archimedes in his shipbuilding schemes.

One day, as another supership was getting its finishing touches, Hieron threw out a challenge to Archimedes, who'd been mumbling away about the input and output forces that the complex pulleys he'd invented gave him.

Hieron said, "You're always going on about that leverage nonsense. I bet it'll take an army of men and oxen to move that new ship of mine. Willing to put your money where your mouth is—an actual experiment?"

Archimedes rose to the bait. When the day arrived, the bearded genius approached the shore. Word of their experiment had gotten out, and would later be described by several Greek writers of note. An excerpt from Plutarch's reportage: "Hieron, entreating him to make good this problem by actual experience, and show some great weight moved by a small engine, Archimedes fixed accordingly on a ship of burden which could not be drawn out of the dock without great labor and many men; and loading her with many passengers and a full freight, sitting himself the while far off, with no great endeavor, but only holding the head of the pulley in his hand and drawing the cords by degrees, he drew the ship in a straight line, as smoothly and evenly as if she had been in the sea. The king was astonished at this."

Athenaeus, another ancient source, told the story differently: "As each part [of the ship] was completed, it was overlaid with tiling made of lead ... This part of the ship, then, was ordered to be launched into the sea, to receive the finishing touches there. But after considerable discussion in regard to the method of pulling it into the water, Archimedes the mechanician alone was able to launch it with the help of a few persons. For by the construction of a windlass he was able to launch a ship of so great proportions in the water."

One way or another, the great ship moved. Archimedes had proven the soundness of his fundamental principle. Now he knew he could move anything—even an entire planet, provided he had another planet to stand on. In that triumphant adrenaline moment, Archimedes crowed: "Give me a place to stand and I will move the earth!"

Besides the feat itself, which must have required an ingenious combination of geared winches, windlasses, and compound pulleys, the power of his words has echoed through the centuries. Generations of speakers, including JFK and a host of other U.S. presidents, have quoted Archimedes. The simple language of his Doric Greek has been puffed up, added to, and misused by many, but the truth of it remains a mathematical certainty.

Archimedes had countless pet projects of his own. He developed an orrery, a contraption with moving parts that showed the sun and the planets in relation to one another. He's also credited with an invention that's had applications as diverse as the iPhone for over two thousand years now. Dubbed the Archimedes screw, it was a hardworking pump that moved quantities of water uphill via a helical screw mounted slantwise inside its cylinder. A hit among farmers for irrigation and miners for draining soggy mineshafts, it also delighted the maritime industry as the world's first bilge pump. It's still in constant use worldwide. Among other uses, it's now employed as a nanodevice in cardiac assist systems, to maintain blood flow in patients undergoing heart surgery.

After studying working levers and an earlier treatise called *Mechanical Problems*, Archimedes wrote *On the Equilibrium of Plane Figures*, the first formal mathematical proof of the lever. This busy man also wrote volumes on mathematical and scientific theory, gravity, the parabola, and buoyancy. Of these books, only a fraction remains. He's considered the founder of hydrostatics, integral calculus, static mechanics, and mathematical physics. Archimedes was the first to prove that the circumference and the diameter of a circle are related by the math constant called pi.

Once upon a time, Archimedes supposedly ran naked through the streets of Syracuse shouting "Eureka!" because while taking a bath, he'd discovered the principle of displacement and the difference between the weight of materials and their density.

In 2006, scientists echoed his cry. That year, through the magic of multispectral digital imaging, an ancient palimpsest revealed three new texts of Archimedes, including *Method of Mechanical Theorems* and *On Floating Bodies*. They'd been hidden for eight centuries in a much-recycled sheepskin parchment that had served as a prayer book, among other things.

While thrilled by the newfound treasures, observers were a bit let down when the discoverers announced their find at a mundane press conference rather than exhibiting that exuberant Archimedean spirit by streaking starkers through the streets, as Archimedes would have.

END GAME FOR THE GEEKIEST GREEK

Some math geeks live and breathe mathematics. Archimedes not only lived and breathed numbers, he may have expired mathematically. In mid-equation, as it were.

Old geniuses get crabby when interrupted. In 212 B.C., this led to a historic "oops" moment—and Archimedes' ad hoc demise.

He had the luck to be born into a noble family in a warm and prosperous walled city, famed for its Olympics-winning thoroughbred horses and its gleaming silver coinage. His Syracuse boasted two natural harbors, a vibrant arts scene, and one of Italy's most beautiful theaters.

Like most mathematical geniuses, Archimedes would have been content to stay in ivory-tower isolation and pursue the problems his fertile brain devised. As the years rolled on, however, his strategically important home increasingly found itself between a rock and a hard place: the military superpower of Carthage on the North African coast and territory-hungry Rome, which had conquered southern Italy and now looked greedily at Sicily.

His younger relative Hieron II, Syracuse's longtime ruler, persuaded him to put aside his pet projects to help him save their home city. Archimedes attacked these problems with alacrity. For Hieron, and later for his son and heir Gelon, the tireless tactician came up with an array of the most amazing armaments. Since Syracuse suffered from the constant threat of sieges from the Romans and Carthaginians, Archimedes' first goal was to further protect the city walls, already thought to be impregnable.

Adapting and inventing a wide spectrum of defensive weapons, Archimedes designed massive catapults that threw stones weighing hundreds of pounds at targets both short-range and long-range. He developed special cranes that swung out over the battlements to drop large timbers on opponents. He pierced the city's walls with specially tapered holes that allowed archers with scorpion crossbows to fire and reload with impunity from inside the thick walls.

Then he turned to offensive weaponry. Archimedes' most terrifying superweapon? His war engine, called "the claw." At the water's edge, upon one of the tall towers of his city's battlements, the inventor erected a giant crane whose beam swiveled vertically and horizontally. At the business end, suspended from chains and dangling over the water, was a counterweighted boom with massive grappling hooks of iron. On the land side of the device, teams of oxen pulled the beam into the air.

When an enemy vessel approached, the claw was lowered to hook onto the ship. This was easier to do than you might think, since Greek and Roman warships had sharp bronze ramming prows at the waterline, protruding a number of feet in front of the vessels. Once the iron claws grabbed on to a ship, they snatched it into the air, then abruptly dropped it, causing the vessel to crash onto the rocks. Like the grisly special effects in a Japanese horror movie, the claw could also shake the vessel so that the men inside were thrown into the water.

A stunned historian who'd seen it in action said, "Some of the vessels fell on their sides, some entirely capsized, while the greater number, when their prows were thus dropped from a height, went under water and filled, throwing all into confusion." Confusion, as in panic and death. After being battle-tested against the claw, surviving Romans called it the biggest fear factor they'd ever encountered in battle.

Although the lavish praise from grateful Syracusans and young King Gelon was nice, Archimedes kept wondering when he'd get to retire from active duty. His kinsman, longtime ruler Hieron II, had died five years ago. And Archimedes was pushing into his seventies, for pity's sake.

The year 212 B.C. arrived; the current siege by General Marcellus and the Romans was now at the two-year mark. The local citizenry needed some cheering up, and everyone thought, what better way to forget civic troubles than with a little libation or two at the Festival of Artemis? Intent on the all-night celebration, the Syracusans eased their watchfulness. Marcellus, who'd noticed a slackly defended stretch of wall, used the opportunity to send a strike force of a thousand men with ladders to scale it. At dawn the Romans broke into the city, soon commencing the slaughter of hung-over Syracusans.

One version of the story has it that deep inside the city, Archimedes was concentrating on a problem, doodling away on one of his theorems. Lost in thought, he paid no attention to the distant ruckus of rape and pillage. When a lone Roman soldier showed up, rudely ordering him to drop what he was doing, Archimedes protested. By now he was a right crabby old genius, so he was pretty testy about it. Anxious to get back to looting with his comrades, the soldier ran the old man through with his sword, trampling Archimedes' equation underfoot.

Upon learning of the old man's death, the besieging general, Marcellus, wept at the loss of this irreplaceable secret weapon he'd wanted to capture for Rome—the beautiful mind of the world's greatest mathematician and weaponry guru. (Abject sorrow didn't stop Marcellus from snatching up all the best sculptures and artwork in Syracuse, however.)

Along with Newton and Einstein, Archimedes is still regarded as one of the three greatest mathematicians of all time.

THE GRINCH WHO STOLE SATURNALIA

Syncretism, the "why reinvent the wheel" tactic used by various religions and cults, has had a long and honorable history. It only made sense for those endeavoring to win converts and supplant another creed to preempt that religion's most beloved traditions. Especially holidays.

In the first few centuries A.D., most of the folks in the Roman Empire could be excused for lumping together those who practiced Judaism with another monotheistic splinter group called the followers of Chrestus, whose activities were at first dismissed as "mischievous superstition." Once everyone had finally caught on to the differences between the long-held beliefs of the Jews and the quirks of the newfangled Christianity bunch, the latter began to grow its numbers in a significant way. In a century or two, the Christian church developed a hierarchy, corporate strategies, and branch offices in various parts of the Roman Empire.

Whenever the Church fathers got together for brainstorming potlucks, they sorted through the best ideas of the average pagan to see what could be recycled to their advantage.

Since time immemorial, the most anticipated festival of the Roman year had been Saturnalia, a week-long brouhaha in December. It had everything: funny masks and costumes, freedom to gamble, license to flirt, extra points for drinking heavily. Even a reversal of status, so that slaves got to lord it over masters for a heavenly day or so.

The solemn opening day ceremony to Saturn, the god who'd reigned during a long-ago golden age, was a pip. Lots of pageantry and splashy animal sacrifices, followed by a meaty banquet open to the public, where everyone, now in casual clothes and pointy red felt caps, gorged themselves and shouted "Long live Saturnalia" until they were hoarse.

An ancient harvest god, Saturn ruled during the golden age of man. At his celebrations, revelry reached Mardi Gras intensity.

The parties that followed were unreal. Some festivities took place in public, where tipsy citizens in costume danced through the streets, visiting neighbors. Shops closed, business ground to a halt, government workers didn't make even a pretense of labor. Household members decorated each home with evergreen boughs, then elected their own "king of Saturnalia." Everyone, slaves included, got extra wine and baked goodies. The gifts poured in: kids received ceramic dolls, the ancient equivalent of Barbies. Grown-ups gave jewelry to close pals and exchanged wax candles, fruit, and incense, all symbols of good luck, with the rest.

But, as the Christian fathers reminded each other, Saturnalia wasn't the only crowd-pleaser. The Mithras bash had success written all over it. The followers of Mithras, a rough, tough, bull-slaughtering, bloodletting, males-only religion that appealed to soldiers and vets, was a growing sect that presented Christianity with some very stiff competition for new members. They held a resplendent year's-end holiday called Sol Invictus, the Festival of the Unconquered Sun. It honored the winter solstice, and its activities included lots of gluttony and gift giving as well.

As they sat and chatted in their early Christian boardroom, the fathers of the one true church grew somewhat gloomy; how could they compete with the festive blandishments of Saturnalia and that Unconquered Sun business? All they had was the birthday of their savior, the exact date of which still provoked bitter quarrels among the faithful.

Finally, one visionary thinker threw a daring suggestion at the group. "Forget this fussing about actual birth dates," he argued. "Let's just borrow the best from the whole shebang, and schedule our annual Christ Mass to fall at the same time."

Everyone gasped, then grew thoughtful. Outrageous! Tricky! But doable. And it was done. By degrees, Christmas adopted many of the trappings and customs of Saturnalia, winter solstice celebrations, and the Unconquered Sun bashes—then added a few twists of its own. As the early Christians well knew, imitation was the sincerest (and most successful) form of flattery on earth.

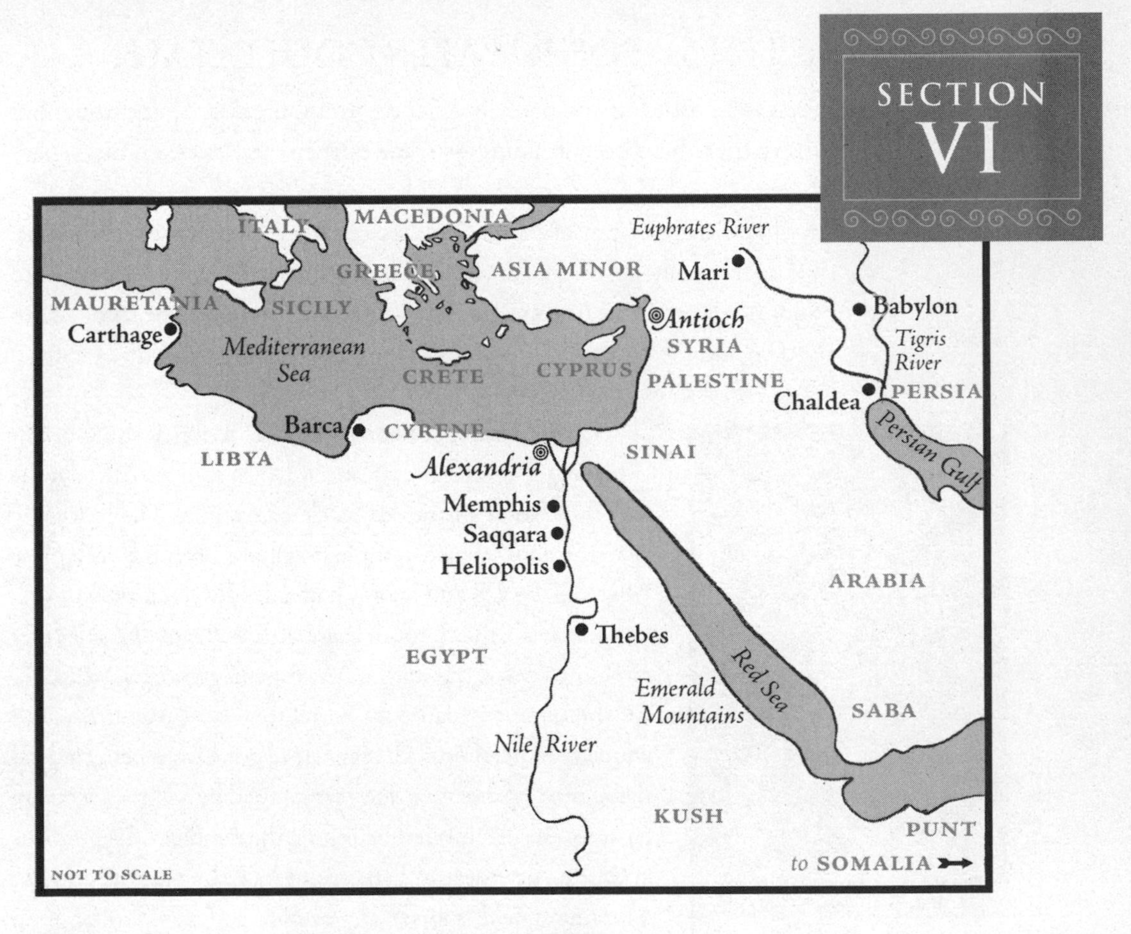

North Africa & Mesopotamia

IF THESE FOUR WALLS COULD TALK

Blame it on bad feng shui or an ill-omened groundbreaking, but throughout history there have been buildings—some of them architectural marvels—where tragedy lurks, and returns to strike again.

That was the destiny of the famed Caesareum on the shores of Alexandria, Egypt, originally begun as a simple harborside altar built by Cleopatra to honor her lover Marc Antony. You know building projects; one thing led to another, and pretty soon Cleo's altar had grown into a temple with a matching set of Antony action figures.

When not busy swallowing pearls to outspend Marc Antony, Queen Cleopatra built things—including an altar to her lover.

On the second of September in 31 B.C., that shiny-bright beginning shattered after Marc and Cleo suffered defeat at the hands of Rome in the sea battle of Actium. The following year, the altar-turned-temple may have been the very place where Cleopatra and an asp had a deadly rendezvous.

Octavian, the victor of that battle and later the sole ruler of Rome, tossed out all the Antony imagery. His makeover for the temple became an homage to his great-uncle and adoptive father, Julius Caesar, who'd gone from being a dead dictator to a protective deity worshiped by sailors. Octavian enlarged the precinct, embellishing the temple with grandiose artwork, porticoes, and amenities including everything but a Dramamine dispenser, the whole affair a symbol of good luck for sea voyagers. Its official name was now the Caesareum.

Marvelous and ostentatious as it was, the temple still lacked a certain *je ne sais quoi*. To jazz up the main entrance, around 13 B.C. Octavian Augustus (now de facto Roman emperor) went shopping. In Heliopolis, south of Cairo, he found a pair of sleek 200-ton red granite obelisks and had them dragged to the temple's front entrance, where they were installed with due ceremony.

Although the obelisks had been made in 1450 B.C. for Tuthmosis III, folklore soon took over. No one in Greco-Roman Alexandria could read hieroglyphs anyway, so the 68-foot-tall obelisks soon collected the sexier handle of "Cleopatra's Needles."

(Flash-forward: one obelisk eventually did a face-plant and was scarfed up in 1877 by the British and installed on the Thames River embankment in London. Its twin in Alexandria stood for centuries, its edges and hieroglyphs getting sandblasted by winds. In 1879, an Egyptian pasha presented it to the Americans, who gleefully set it up in New York's Central Park.)

As the years passed, the emperor kept busy with empire expansion. As a young man, he'd achieved supreme power in the Roman Republic in part by calling himself "son of a god, the deified Julius Caesar." Nevertheless, he frowned on being worshiped as a deity himself. He died in A.D. 14 but wasn't even out of rigor mortis before his widow, Livia, pressured the Roman Senate to declare him a god.

New temples immediately sprang up for official Caesar worship. Myriad others, including the temple in Alexandria, were simply rededicated. The Caesareum, now renamed the Sebasteum (meaning "venerable one"), continued as a center for the worship of two emperors.

Centuries rolled by. As Christianity took deep root in the area, the Caesareum-Sebasteum lost its pagan identity and was repurposed as the Cathedral of Alexandria. Over time, rifts widened between the pagan, Jewish, and Christian communities of the city. Religious fanaticism and riots became more frequent. In A.D. 414, matters came to a head when the civic leader Orestes and Bishop Cyril locked horns over control of Alexandria's worshipers, by then the largest Christian community in the world.

Alexandria's most recognized figure of the day was neither a civic bigwig nor a religious leader. She was a philosopher, inventor, and teacher of sciences whose pupils and allies included Orestes and other Christians of note.

Her name? Hypatia, daughter of Theon the astronomer. Father and daughter had different philosophies. A professor at the Great Library and Museum, Theon was drawn to astrology and more mystical ways of seeing the world. A Neoplatonist, Hypatia had a more rational approach. Her genius at math, geometry, and astronomy won praise everywhere; her writings on conic sections and her new design for an astrolabe had eggheads atwitter. Her fan base extended around the Mediterranean; letters addressed simply to "The Philosopher" reached her.

Although a moderate, Hypatia was outspoken. High-profile. And a single woman of mature years. An ideal target for religious extremists, in other words.

On a spring day in 415, Hypatia was driving her chariot by the shore when a mob of fanatical monks attacked and killed her in the cathedral, still known as the Caesareum by locals. The tragedy shook her city to the core. One of Hypatia's students, a legal counselor called Socrates Scholasticus, described her murder in graphic detail: "Certain rash and heady cockbrains whose guide and captain was Peter, a reader of Caesarium Church, watched Hypatia coming home, pulled her from her chariot, hauled her into the church, stripped her naked, and rended the skin and flesh of her body with sharp shells, until the breath went out of her body. Then they quartered her, brought the quarters to a place outside the city called Cinaron and burned them into ashes.

This heinous offence was no small blemish both to Cyril and the Church of Alexandria."

No one was brought to justice for these loathsome crimes. No one pointed the finger at Bishop Cyril, even though he'd publicly vowed to rid the city of Neoplatonist "heretics," as he called them. Instead, he rose to immense power in Alexandria, ultimately being made a saint.

The Caesareum/cathedral, that mockery of religious tolerance, stood for five hundred years more. Today only a few of its forlorn stones still linger in Alexandria. Perhaps this is just as well.

HOT ROCKS, PEARLS OF WISDOM

Gem lore lasts forever, as durable as diamonds. The Greek word for diamond was *adamas* (our *adamant*), meaning "unconquerable force." Besides the real article, the Greeks labeled any number of minerals (iron pyrites, for one) as *adamas*. Two thousand years ago, legend had it that diamonds couldn't be burned by fire—but could be shattered with the help of fresh, warm goat's blood.

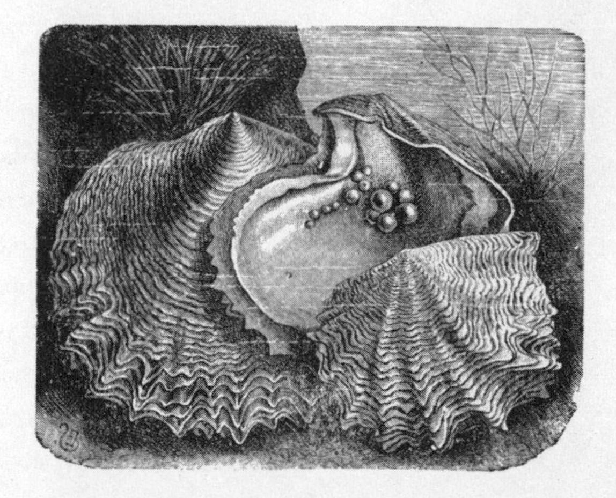

Back then, even though there were plenty of heinously rich people willing to pay anything to sport some bling, there weren't that many kinds of sparkling gems available. The diamonds of yore would not be called a girl's best friend today. Small, off-color, often flawed, they were sometimes confused with rock crystal. Even during the long and prosperous reign of the Roman emperor Octavian Augustus, a paltry amount of quality diamonds came out of India, and artisans had no tools hard enough to cut them into glittering facets. Indeed, Greek and Roman jewelers were grateful to obtain diamond splinters, which they used to tip their tools and cut other gems.

Emeralds, therefore, took center stage as the most sought after of true gemstones. After the defeat and suicide of Cleopatra in 30 B.C., Romans promptly began exploiting Egypt's mineral wealth, tapping into the emeralds mined from Mons Smaragdus (Emerald Mountains) near the Red Sea.

As soon as emeralds hit the Rome fashion scene, the nouveau riche did their best to vulgarize their use. For example, the patrician Lollius family set the tone when Marcus, the paterfamilias, plundered the eastern provinces he'd been given to govern and took part of his loot in gems before getting called on the carpet. Although he was forced into early retirement via mandatory suicide by poison, he left his family well fixed.

Two emperors later, his flashy granddaughter Lollia Paulina caught the eye of Emperor Caligula at a party. More accurately, she blinded him with her person, being covered head to foot with interlaced pearls and emeralds. In case anyone inquired as to their worth, Lollia carried the official paperwork in her capacious bosom. Caligula, always on the lookout for new income sources and girlfriends, did inquire. When he got the answer, 40 million sesterces, he soon proposed marriage, a relationship that lasted about as long as the average twenty-first-century celebrity hookup.

More modest fashionistas could also find emeralds or "emeralds" in their price range; Pliny listed twelve varieties, only a few of which were the genuine article. Various stones, including the lovely pale green peridot, wore the "emerald" label and price tag, so the market was more caveat than emptor. Funeral portrait art from Greco-Roman Egypt during the first through third centuries A.D. shows a variety of damsels young and old resplendent in gold necklaces and earrings heavy with green gems.

Deep-pocketed buyers for other precious gems were limited to imported rubies, sapphires, and beryls. Beyond that, the most popular semiprecious choices were opals, amber, onyx, sardonyx, garnets, lapis lazuli, rock crystal, and agates or fluorspar. The last item, called myrrhine by the Romans, was carved into glorious bowls and vessels; Emperor Nero once laid down a cool million sesterces for one.

Where there's a gem, or even a "gem," there's a fabulous fake—as true long ago as it is today. When emeralds became the hot rocks, an entire industry sprang up to produce sparkling imitations. Other gems and semiprecious stones also lent themselves to fakery. Forgers were already onto the foil-backing trick; they also used heat, oils, dyes, and other methods to turn dross into ersatz beryls, sapphires, opals, and turquoise. Some of the best faux gem manufacture may have been done by rogue alchemists, whose deep knowledge of chemicals and reactions would have made this a lucrative sideline.

Jewelry wasn't merely a female addiction. Men wore seal rings and signets, sometimes one per finger, driving demand for gemstones and semiprecious stones. Sards and carnelians were particularly desirable because the intaglios carved into them made a clean impression in clay or on papyrus. Ring "signatures" were used to seal containers, sign documents, and serve a variety of legal and financial purposes. Patricians weren't the only ring wearers, either; freedmen, who carried out the lion's share of day-to-day business, proudly wore them as well.

Gemstones had further utility beyond beauty. Malachite conferred protection and brought good luck. Sandastros, thought to be aventurine quartz, was a must-have for astrologers. Its inclusions glittered like golden stars, resembling certain constellations. Amethyst had a loyal following due to its hoped-for abilities to prevent drunkenness.

Although pearls weren't gemstones at all, their glossy, glowing beauty made them objects of desire anyway. Stories about fantastic pearls added to their mystique. Cleopatra supposedly swallowed the largest pearl in the world in order to impress her lover Marc Antony with her wealth.

Cleo loved to kid around. They both did. She'd made Antony a bet that she could outdo his generosity by spending ten million sesterces on a single banquet. When the food arrived, Marc dug in—but only a goblet was set before Cleopatra. Instead of wine, it contained vinegar, to which she added one of her pearl earrings. In ancient times, there was widespread belief that vinegar could melt even the biggest pearl. Cleo, of course, knew better. Marc probably did, too.

She downed the pearl in vinegar and won the bet. He, on the other hand, spent several anxious nights waiting for her to pass the darned thing. Marc always had cash flow problems. A vulgar lunkhead but a practical sort, he knew that what went in whole often came out the same way. Once tidied up, a pearl of that size could serve them well.

THE SCIENCE OF NANO-ARTISTRY

How on earth did craftsmen of long ago achieve their incredible standard of artistry and detail in the signature rings, coins, and models they made? Traditionally, we've been told that their societies had no magnification tools or eyeglasses. Was that really so? Given the sheer quantity and quality of the nano-art that has survived, to say nothing of all the items described but now lost, it's hard to believe there were enough astigmatic jewelers and artisans with 20/15 eyesight to fill the demand for such top-notch work.

In Pliny's encyclopedia of oddities and natural wonders, for example, he described minute works of art which he himself doubted could have been done without mechanical (or magical) means of magnification. An excerpt: "Cicero records that a parchment copy of Homer's *Iliad* was kept in a nutshell. Callicrates used to make such small ivory models of ants and other creatures that to anyone else their parts were invisible. A certain Myrmedides was renowned as a model-maker for his four-horsed chariot carved from ivory, so small that a fly's wings could cover it. He also made a ship so tiny that a small bee could hide it with its wings."

Here and there in the works of Athenaeus, Aulus Gellius, Pausanias, and other Greek and Roman writers are hints about the use of water-filled glass globes. Scribes could also have used such devices for magnifying the pages they worked on. Even a halfway alert specialist would have noticed the ability of a single drop of water to magnify.

From water to rock crystal was a logical step. Well known and prized in ancient times, it was also called the "rainbow stone" for its prismatic qualities in refracting light. Most rock crystal came by way of India or the Alps. Its most intriguing characteristics: transparency and its two surfaces, one flat and another that could be convex or concave. Rock crystals also took a high polish. Starting to sound like the lens in your prescription eyeglasses? It should.

Lenses made of rock crystal have been found at a growing number of archaeological sites: at Troy, on Crete, at Carthage, and at the Pompeiian house of a man identified as an engraver. In his survey of world museums with antiquities collections, researcher-author Robert Temple in his book *The Crystal Sun* identified 450 such lenses and magnification aids, many heretofore enigmatic or ignored as to purpose.

One hoard excavated in Ephesus (Asia Minor) contained concave lenses only, which were referred to as optical instruments by the principals of the dig. On Crete, a highly polished plano-convex rock crystal lens, found in 1983 in a sacred cave, possessed excellent clarity and a useful magnification of seven times. It was far from an anomaly.

Some convex rock crystals also had the ability to gather light as they magnified. These light condensers were noted by Temple, who photographed their simultaneous magnification and lighting of a page.

Modern researchers have long pooh-poohed the old wheeze about the Mr. Magoo eyesight of Emperor Nero and his supposed use of an emerald to soothe or improve his vision at the gladiatorial games. Emeralds were at that time the most expensive gems on the planet. But hey, he was the emperor.

A more plausible and fascinating scenario has been proposed by Temple. As noted earlier, for centuries before Nero, ancient entrepreneurs had engaged in the profitable scam of gem faking. They would soak rock crystal in a sulfur bath to tint it emerald green. This was referred to as "baptizing the stone"; it may have involved the use of fire also. Thus if Nero's "emerald" was actually green-tinted rock crystal, a plano-convex specimen would have the power to improve his nearsightedness markedly.

Another commonsense clue that points toward ancient knowledge of the optical properties of rock crystal (and possibly other materials) comes from the fieldwork of archaeologists themselves. Those who specialize in cuneiform tablets or ancient coinage routinely use magnifying devices to study the details. It would seem reasonable that similar visual aids would have been required to create those micro-objects, minuscule symbols, and nanoletters in the first place.

Take the existence of microletters on coinage. Hidden in the details of hair or jewelry on coins, modern investigators have found names and letters, placed there by engravers or die cutters. Although they may have done so to "sign" the piece, the most likely reason would be to outwit numismatic forgers.

Another ancient use for lenses made of rock crystal or other materials was to concentrate the sun's rays, something we've all done as Scouts to start a fire—or to annoy friends who've fallen asleep while sunbathing. Roman doctors apparently employed such lenses to cauterize wounds. The vestal virgins of Rome also kept their sacred flames to Vesta going with rock crystals.

These lenses must have been common objects in the ancient Greek world as well. One piquant piece of evidence: the comedy called *The Clouds*, written by Athens' favorite comic playwright, Aristophanes. In it, a rascally protagonist named Strepsiades, in debt and longing to outwit creditors, studies the art of sophisticated double-talk from a character who's a send-up of the real-life Socrates. He obliges by engaging Strepsiades in Socratic questions and answers.

Socrates asks, "Suppose you lost a lawsuit for five talents; how to evade payment?"

Strepsiades has the clever idea of making his lawsuit "disappear" by melting the wax off the wax tablets it was written on. He tells Socrates, "At the druggist's, you've seen the fine transparent stone with which fires are kindled?"

Socrates answers, "You mean crystal lens?"

"Yes," says Strepsiades. "I'd get one, and when the clerk was entering my sentence, I'd stand a ways off, and melt the writing away."

"Neat, by the Graces," responds Socrates.

The audience roared. If such transparent stones hadn't been everyday items, the humor of that exchange would have vanished like melting wax.

SHEDDING LIGHT ON NAUTICAL TRAVEL

We tend to think of lighthouses as structures that give off audible and visible warning signals round the clock. Ancient folks, however, didn't see a crying need for such a thing at first. To begin with, boats of all sizes tried to move by day only; any captain with a lick of sanity got his vessel and crew ashore before dark. Only when ghastly weather or a similar calamity arose did boats move—or even anchor—at night.

In fact, the first lighthouse in the Greco-Roman world to shine after sunset was built in A.D. 46, erected by Rome's favorite madman, Emperor Caligula. He placed it on the shores of the present city of Boulogne, France, to commemorate his nonexistent "victory" over the god Neptune on the shores of the Atlantic—one of his many colorful delusions. Nevertheless, the slender 200-foot structure was one delusion with practical benefit to the mariners struggling to navigate the treacherous English Channel.

Stoke those flames, kid—this temple's open 24/7 and needs to be visible day and night.

The great lighthouse at Alexandria, called Pharos after the island it sat on, was of much greater antiquity and fame. About 280 B.C. it was erected by the first Ptolemy and his architect Sostratos at a skyscraper cost of some 800 talents. Although topped by a glorious statue of Isis Pharia that guided ships into the harbor by day, the 45-story-high landmark didn't get its after-dark firepower until sometime between A.D. 41 and 65. That was when permanent fires, using wood impregnated with tar, resin, and asphalt, were lit in the top story, giving those distressed souls still at sea a comforting glow that could be seen for 30 miles or more. Some accounts also claim that the Pharos had a lens of glass or crystal that magnified the fiery impact of the flames.

By degrees other lighthouses came on line and began to function day and night around the Mediterranean. One was built at Ostia by Emperor Claudius at the same time he enlarged the harbor and constructed breakwaters to serve Rome. Always a keen recycler, Claudius used a supership to import one of those giant obelisks of which Egypt had such an abundance—then sank the vessel and had a lofty lighthouse tower erected on the pilings.

During the reign of Emperor Trajan, a pair of 80-foot octagon-shaped lighthouses were built near Dover, England. One of them, now located within the precincts of Dover Castle, still boasts half its original stature.

By A.D. 200, as many as thirty lighthouses had been built along coasts from southern Italy to northern Spain. The one at Coruna, Spain, modeled after the Alexandrian giant, still stands on its sheer cliff above the harbor. Most lighthouses were called pharos, a word that became synonymous with the original in Alexandria.

Prior to this, how did mariners navigate the treacherous coastlines, tricky currents, blustery winds, and impetuous weather of the Black Sea and Mediterranean? In places there were watchtowers on cliffs, sometimes manned, in order to warn ships off with torches.

Thanks to archaeologist E. C. Semple, who made it her life's work to study such matters, it's clear there was another surprising source of illumination along many coasts: the temples to Apollo, Venus, and Neptune. In these calm sanctuaries, priests and priestesses regularly kept fires going. There may have been as many as two hundred such structures, some of them simple shrines, others as elaborate as the beautiful, many-columned temple at Sounion, Greece. Their distribution often matched sites of particular difficulty or danger for ships, such as the tip of the Sinai Peninsula in the Red Sea, much dreaded by mariners rounding it because of its combative currents and winds.

The skippers of Greek and Roman ships had more problems than the lack of night-lighted lighthouses, however. A majority of vessels, more especially the ubiquitous trireme warships, had urgent motives besides darkness to seek the shore. Most relied on rowing power rather than sails. Thus their interiors were jammed with sweating humans rather than galleys, bunks, toilets, and equipment. Every oar-powered vessel afloat had to haul out during daylight hours in order to cook, feed their crews, catch some sleep, and attend to more delicate matters.

Another little problem: given their emphasis on speed and lack of room, warships did not keep their auxiliary sails aboard; they too were stored on shore, especially during battles. Naval engagements were clearly daytime affairs. To be able to locate their own sail stashes ashore with any reliable frequency, trireme commanders must have been a keen-eyed lot. Did naval opponents ever cross paths onshore? What were the rules of engagement while searching for one's sails? What happened if your land support didn't show up? It's the quotidian minor mysteries like these that keep historians hooked.

Alchemists chose this symbol to represent spring

ROMANCING THE PHILOSOPHER'S STONE

Alchemists, the secretive, slightly wild-eyed forerunners of modern chemists, thought of themselves as trailblazers decoding the mystical enigmas of the universe.

Alchemy got rolling as a pseudoscience based on a misguided idea from dear old Aristotle (among others), who dredged it up while reading. According to this theory, every bit of prime matter was made of the same four essences or ingredients—earth, air, fire, and water—in some combination. The recipe—and therefore the end product—could be changed by applying heat, cold, dryness, or wetness. Moreover, it was an article of faith that a master cookbook called the philosophers' stone existed. In ancient times, all alchemists were in competition to chase down those recipes.

The word *chemeia*, the Greek name for chemistry, meant "to change matter." For many, the quest was metaphysical, noble. For other alchemists, the motivation was the Mammon-esque prospect of transmuting useful but dull metals such as lead into more perfect and cashable ones, particularly gold.

The specific goals of these early protochemists contained threads of animist belief. They were convinced that every object possessed some form of life. Marble could regenerate itself. Metals could grow. Adepts claimed that mercury was the embryonic womb in which other metals could be gestated. These notions, carried to mad extremes by the alchemists, came from the likes of Thales, Heraclitus, and other early thinkers who'd embraced the idea of a living, interconnected cosmos.

The first alchemist to get his "best secrets" recipe book to market was an Alexandrian Greek named Bolos Democritos. Its title: *Physika*. To him, the most important characteristic of metal was its color. To transmute it, he insisted, there had to be a certain sequence of color changes for a successful result. His secret was to transform the substance by stages, first to black, then white, followed by iridescent, yellow, purple and finally to red, the desired color. (One of his chapters was called "The Making of Gems," so you can see where he was going with this color idea.)

The alchemy symbol for summer

He and his competitors carried out this transmutation chore, and others like it, via laborious, noisome, often perilous operations involving red-hot furnaces and vats of putrefying materials. Their methodology had grandiose names, from albification (turning stuff white) to impastation (letting stuff putrefy and thicken until it turned black). Many of them relied on sympathetic magic, the belief that like produces like or "as above, so below." For instance, the Greeks were convinced that if a person suffering from jaundice looked into the large golden eyes of a stone-curlew bird and it stared steadily back at him, he would be cured. In alchemy, getting a metal to turn the color of gold clearly meant you were halfway to the real McCoy already.

These bold hazmat investigators messed about with hundreds of substances, including sulfur, vinegar, niter, vitriol, salt, and verdigris. It must have been an exhilarating subculture. There was nothing like putting in a hard day's alchemy, turning stannic chloride into butter of tin. Or burning zinc to make "philosophers' wool," then excitedly blogging to pals and competitors about the results in language cloaked in the most outlandish metaphor and symbolism.

The symbol for fall

Alchemists got almost as big a bang out of creating fanciful names for things as they did for blowing them up. Mercury, for example, was variously called the seed of the dragon, the divine dew, Scythian water, the milk of the black cow, and the ever-fugitive.

Given the thickets of mystical flimflammery to pore through in long-ago alchemy texts, it can be tough to pinpoint the useful material. Although their "science" was more often the chanting of magical formulae than the crunching of empirical data, alchemists did lay the groundwork for modern chemistry. Through their pursuits, alchemists made serious inroads into ore extraction, metallurgy, the making of alloys, chemical reactions, and pharmacology, which led to unexpected technological advances and serendipitous scientific breakthroughs.

The most successful and best-documented alchemist of the period was a woman we know simply as Maria of Alexandria, Egypt. Thanks to the goofy language and paranoid secrecy of alchemists, her biographical particulars are shrouded in contradiction and mystery. Variously called Maria Profetissa, the sister of Moses, and Maria la Judia, she's thought to have been a contemporary of Archimedes, but her dates could fall anywhere from the second century B.C. to the third century A.D.

A native of Alexandria, Egypt's scientific epicenter, she lived in a city teeming with intellectual excitement. It was also a hotbed of technology and engineering advances, rich in artisans who skillfully developed the fields of glassblowing and metallurgy in particular.

The symbol for winter

Maria may have come from a family tradition of such artisans. In any event, she succeeded in inventing and perfecting a number of key instruments for the pursuit of alchemy. Her brainchildren included a heating apparatus called the kerotakis that used the principle of reflux cooling to treat metallic objects with vapors. She also came up with the tribikos, a still with three glass spouts—an excellent device for distillation, with wider applications for perfume making and chemical analysis. She's often named as the author of a now-vanished work called *Dialogue of Maria and Aros on the Magistery of Hermes*. Her name is also linked to the discovery of hydrochloric acid and the study of sulfur compounds. Some sources also gave Maria credit for creating niello, the black compound used for metalwork inlays. In addition, her name as inventor of the double boiler would be carried forward in time. First given the homely name of Mary's bath, which passed into Latin as *balneum mariae*, today's double boiler is still called *bain-marie* by French cooks and *bano-maria* by Spanish ones.

Not all alchemists were as honorable and pragmatic as Maria. Predictably, some (such as Bolos Democritos, evidently) turned to shabbier pursuits, including tingeing metals and other substances to resemble gold or precious gems. Dozens of papyri with such formulas have survived from ancient scalawags. One for-instance came with its own "buyer beware" postscript: "To manufacture silver, purify white tin four times, melt six parts of this and one mina of white Galatian copper; rub off and make what you wish. It will be silver of the first quality, although artisans may notice something peculiar about it because it's formed by the procedure mentioned."

Another example: "Preparation of a [fake] pearl: etch crystal in the urine of an uncorrupted [that is, virginal] youth mixed with alum; then dip it in mercury and a woman's breast milk."

Other pioneers maintained a purist attitude, interested only in the mystical aspects of alchemy. They referred to the more practical alchemists as "puffers," a description of the bellows used to stoke alchemists' furnaces. As time would reveal, it was experimental practitioners like Maria who acted as midwives, through alchemy enabling chemistry to be born.

LIVING LARGE, LIVING UNOBTRUSIVELY

Of all the wisdom-seeking philosophies and schools of thought from the Greek and Roman eras, none has gotten more wrong-headed dismissals than Epicureanism.

The founder himself, a gentle, plainspoken man, often said, "We must laugh and philosophize at the same time." His collective in Athens attracted scores of followers, including women, foreigners, and slaves. Epicurus embraced everyone—sometimes literally. No one was turned away from his school, where he expounded on his simple road rules for life: "Nothing to fear in the gods. Nothing to feel in death. Good can be attained. Evil can be endured." Predictably, these radical ideas got him into very hot water in misogynistic Athens.

So where in Hades is all that "wine, women, and song" so often linked with Epicureanism?

It could be found at the very different philosophical headquarters of Aristippus, an ardent pursuer of hedonism. A disciple of Socrates a century before Epicurus, this aristocrat was born into wealth and lived part of the year in Athens, the rest in places such as Sicily or Cyrene, his Greek hometown on the North African coast.

In Athens, he participated in a time-share; instead of three weeks at a condo, he split a courtesan's time with another philosopher. His pricey companion, a dishy young woman named Lais, evidently had a thing for philosopher johns.

Although he admired his mentor Socrates' no-fee teachings about excellence and virtue, Aristippus preferred to live large by charging large. His fees were astronomical. One father, when told his son's schooling would cost 500 drachmas, howled, "For that much loot, I can buy a slave!" An unruffled Aristippus replied, "Then do so, and you'll have two."

In the flesh, Epicurus probably wore a smile. He often said, "We must laugh and philosophize at the same time."

Hoping to impress Socrates, Aristippus initially sent his teacher a portion of his fees; the Athenian returned it with a note saying the money annoyed him, and moreover, he'd gotten a supernatural sign not to take it.

Aristippus had a talent for sucking up to royalty. As a consequence, he often made fiscally pleasing stays at the court of Dionysius, ruler of Sicily. Detractors called him "the king's poodle," but such comments slid off him like water off a duck's back. Eventually he established a plush school in Cyrene, the wealthy city-state in North Africa, overlooking the warm blue seas of the Mediterranean.

Despite his lavish habits, Aristippus taught his own daughter philosophy and supposedly raised her to despise excess. Before he died in 356 B.C., he warned, "Don't set a value on anything you can do without." Fine words from a man who probably left her nothing tangible to inherit. Nevertheless, Arete expanded and improved on her dad's rather shallow teachings.

Fifty years after Arete began teaching in Cyrene, Epicurus traveled from his boyhood home on the island of Samos to establish two schools of philosophy in nearby Asia Minor at Mitylene and Lampsacus. Five years later, he traveled to Athens to start his school, which would be called the Garden.

It was a forlorn period for Athens, whose golden age was now history. Would-be conquerer and world-class bully Demetrius of Macedon twice battered Athens, then waged a brutal siege. Surrounded by hostile troops, the city began to starve.

Those huddled in Epicurus' garden included a number of non-Athenian women, the independent, educated courtesans called heterae. A hetera named Leontium happened to know the besieger's current sweetheart and managed to establish contact. Working behind the scenes, she scored enough beans and supplies to keep the Epicurean community from starving. A letter from Epicurus to Leontium gives a sense of the man and the family feeling among his flock: "O Lord Apollo, my dear little Leontium, with what tumultuous applause we were inspired when we read your letter."

A prolific writer, with more than three hundred books to his credit, Epicurus was also an avid social reformer. He encouraged ordinary men and women to aspire to higher education and longed to liberate thinkers from the superstitious elements of religion. He had nothing but scorn for the la-di-dah elitism of the aristocratic crowd, and the prejudice toward any non-Athenians that Plato, Aristotle, and others advocated.

His bold positions and fearless advocating for those marginalized by society drew political fire, scathing remarks, and nasty slanders. Other philosophers fumed about his dalliances with prostitutes, muttered darkly about his "pleasures," and wrote erroneously about his beliefs. Much of the inaccurate reporting began here. Later writers also did a good job of hopelessly conflating his teachings with those of Aristippus.

Epicurus had the last laugh. The schools he founded in Asia Minor, on the Greek mainland, in the Middle East, in Italy, and elsewhere endured for more than seven hundred years. At them, his birthday was celebrated with lively joy each February.

From today's perspective, what were the greatest contributions made by Epicurus? Although the tenets of Epicureanism did not find an ardent following in later centuries, through the preservation of his writings via Lucretius and his numerous disciples, this kindly man transmitted some of the most important scientific ideas of his day—most critically, the atomic theories of Democritus and Leucippus.

NATURALS AT PHILOSOPHY

What did Greeks talk about at those all-male drinking parties? One likely topic: batting around the highly speculative idea of female intellectual capacity and women's fitness to philosophize.

Plenty of men roared with dismay or laughter at such notions, but philosophy was a natural outlet for women. Its teachings offered a way to cope in a world filled with disturbing possibilities: death in childbirth, plague, war, shipwreck, rape, slavery, dishonor. Usually married by age fifteen, Greek women spent their fertile years in nearly continual jeopardy. Despite these odds, a number of women possessed sufficient wit, vitality, and luck to survive such hazards—and make more use of their brains in the bargain.

The biggest success story? Arete, who made a Greek city in North Africa the epicenter of philosophic enlightenment for like-minded students. She was the precocious daughter of the hedonist philosopher Aristippus, who'd studied with Socrates but preferred the contemplation of pleasure—and its fulfillment—to intellectual heavy lifting.

Countless women with intellectual curiosity defied their long-ago societies. Many opted to philosophize, both learning and teaching it.

That wasn't Arete's cup of tea; under her leadership, the Cyrenaic school took a more moderate direction. Little is known of her teachings, but a quote attributed to her reveals an unusual perspective for that time and place: "I dream of a world where there are neither masters nor slaves, where everyone is as free from worry as Socrates."

Her school was headquartered in Cyrene, a gorgeously templed Greek city-state with a population of a hundred thousand on the North African coast. In stark contrast to the arid, treeless terrain of what is now Libya, twenty-five hundred years ago the Cyrenians enjoyed a milder climate with ample rainfall, surrounded by lush pastures where fine thoroughbreds and roses prized for their perfume were raised.

Arete taught for three decades, writing forty books on topics as diverse as education and war. Her large circle of students included her own son. Named Aristippus for his grandfather, he succeeded his mother, although he had his own cross to bear, being stuck with the nickname "the Mother-Taught."

When Arete died, her community showed its esteem. Her tomb carved from limestone was adorned with this fulsome inscription: *Here lies noble Arete, the light of Hellas, who possessed the beauty of Helen, the virtue of Thirma, the pen of Aristippus, the soul of Socrates, and the tongue of Homer.*

Centuries before Arete, however, Greek women had already argued or wheedled their way into most of the major philosophical schools of thought.

Plato's mom Perictione, for example. Although her son was light-years from liberated, she won modest repute as a Pythagorean—just not from him. As head of the prestigious Academy in Athens, Plato did accept at least two female students, however grudgingly. In order to attend his lectures on the mastery of Socratic method, however, Axiothea of Phlius and Lasthenia of Mantinea were obliged to cross-dress as male philosophers in training.

On the other hand, Plato's adored guru Socrates genuinely loved women and their intellectual gifts as well as their physical ones. Socrates maintained lifelong friendships with real women whom he learned from and taught, including eloquent Aspasia, longtime companion of Pericles, and more legendary ones, such as Diotima, the priestess of Mantinea, mentioned in Plato's Dialogues.

The Pythagorean school had begun much earlier than the era of Plato's mom. Located on the southern coast of Italy, the commune welcomed women as students. At fifty-six, its founder, Pythagoras, paused long enough to marry a youthful local named Theano, who birthed five children and scribbled the occasional essay while helping to run the school and keep their community in competent order.

Daughter Damo was the apple of her father's eye, a quick-witted pupil who became one of his teachers. She wrote on mathematical figures. It was to her that Pythagoras entrusted his three books of secret writings, asking her never to release them outside the family.

After her father's death, political unrest among the city-states in southern Italy made life difficult for Damo, her husband, and their son. They fled to Athens. Rabid fans of Pythagoras offered her huge sums for his books; she turned all of them down. With the help of Philolaus and other Pythagoreans, she apparently was able to publish her father's writings on geometry. To her daughter Bitale she left the secret writings, their fate and final disposition unknown.

Meanwhile, intrigued by Pythagorean philosophy and its secretive aura, Plato persuaded Philolaus to write the first tell-all about his guru by paying him the stupefying sum of 100 minae. The death of Pythagoras, who'd become a mythical figure even when alive, did little to quench interest in his philosophic way of life. The school survived for centuries after Pythagoras; in the first century A.D., a new, more mystical wave called neo-Pythagoreanism took hold.

So did a rival retread called Neoplatonism. One of its later teachers, famous for the saddest of reasons, was Hypatia of Alexandria. Brilliant at mathematics and astronomy, this far from radical teacher-philosopher fell victim to mob violence in 415 A.D.

Women—whether of noble rank, dubious reputation, or slave status—were warmly welcome at the Garden, the communal school of Epicurus established in Athens. An accepting and loving teacher, Epicurus brought people from all walks and ranks of life into his circle. His beliefs made good sense to many women. So popular was his "live unobtrusively" school of thought that he opened branch schools around the Mediterranean, many thriving for centuries after his death in 323 B.C.

Two of his philosophical standouts lived in Asia Minor: Themista and her husband, Leonteus. Epicurus pioneered distance learning, keeping up a lively correspondence with a multitude of pupils, including her. A charming excerpt survives in a letter to Themista: "I am quite ready, if you do not come to see me, to spin thrice on my own axis and be propelled to any place that you agree upon." Not every female egghead had famous philosophers writing cheeky letters and spinning like tops, but Themista did.

More than sixty-five high-powered women achieved recognition as philosophers in Greco-Roman times—plus countless others who went unrecorded. From today's perspective, such scanty representation may seem like tokenism. Consider the times they lived in, however. Women had enormous barriers to overcome: legal standing just a cut above that of slaves, lack of access to education, and societal pressure on them to lead private domestic lives. That so many dared, that so many did, is remarkable.

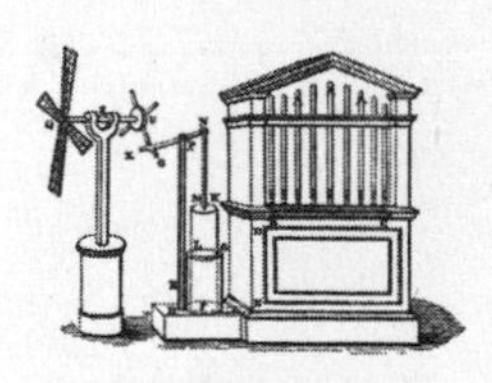

Despite harnessing auxiliary wind power, this organ flunked the sizzle test and failed to excite early adopters.

FEW TAKERS FOR GREEN GOODS

Romans (but more especially the Greeks) were surrounded by H_2O, spending much of their lives on it or near it. Their reverence for fresh water had them worshiping a variety of water gods, river nymphs, and prophetic pond goddesses. Since they had to sail, like it or not, between islands and mainlands with some regularity, they also threw fervent prayers at big-time sea deities Poseidon and his Roman equivalent, Neptune.

Ancient scientists and farmers alike were equally impressed with the sun, calling it the god Sol in Latin or Helios in Greek, and with the winds, identifying and naming eight different directions, with a minor deity for each.

Despite the attention paid to the worship of renewable-energy gods, nobody felt an urgent need to harness wind, water, or solar power. Then again, they weren't hooked on oil, coal, and natural gas the way we are.

Around 270 B.C., at a barber shop in the lively Greek city of Alexandria, Egypt, a canny kid named Ctesibius did cook up the concepts of hydraulics and pneumatics. What's more, he made a musical instrument that apparently worked by harnessing both principles. He'd been fiddling around with pulleys, a lead ball on a cord, and pipes in order to hang a mirror in the family barber shop, when he discovered a useful principle: air, when compressed, made rude noises. Later in his workshop, Ctesibius constructed a gizmo he dubbed the hydro-aulos or hydraulis, inexplicably translated ever since as "water organ," although we think that the aulos, a reed instrument, sounded more like a clarinet or an oboe.

In spite of the geeky name, its popularity grew. By the first century A.D., the hydraulis was hot, the fashionable way to hear loud, lively music at weddings, theater openings, and ribbon cuttings. Gladiatorial arenas everywhere hired their own keyboardists (often depicted as female in ancient art), who cranked out tunes for the crowd along the lines of "Take Me Out to the Bloodbath." Emperor Nero, that musical artiste manqué, went nuts for the hydraulis, playing concerts to captive audiences until A.D. 68, when his fingers were thankfully stilled forever.

Another terrific tinkerer in the same city, Heron by name, admired the hydraulis and thought he might be able to improve on it. Using rods and pistons, he attached the mechanism to a small four-armed windmill, transforming it into a wind-powered organ. Out of the hydraulis, he'd created the aerolis, the first wind-powered Wurlitzer. Even with Heron's name on it, this breezy innovation went over like a lead balloon. As the inventor admitted, the prototype was awkward, since early adopters had to drag the thing around to catch the prevailing winds.

(In more recent times, after finding a squashed but nearly complete hydraulis in a Roman firehouse in what is now Hungary, several scientists, including a Florida professor named Eugene Szonntagh, reconstructed the instrument. From this specimen, it's become clear that the hydraulis used both water and an air regulator called a pnigeus.)

Undeterred by his aerolis flop, Heron whipped up more items run by renewable energy. His solar-powered fountain, for example. As the sun shone on it, the fountain trickled beautifully. A handful of Romans commissioned his fountains in order to sneer at their neighbors' solar-free gardens, but that was about it.

After creating a line of novelty toys that ran on water power, including bronze birds that whistled, a satyr that poured water from a wineskin into a basin, and a cunning scene where a metal archer fired an arrow while a dragon hissed, Heron turned to something more businesslike. He invented a double-siphoned force pump—and holy Sol Invictus, the dogged firefighters of Alexandria, Rome, and myriad other flammable cities finally had a piece of machinery to fight urban fires in a meaningful way.

The prototype was a bit hinky. Heron himself called it "tedious" because it would squirt water to the height needed, just not necessarily in the correct direction—requiring the entire machine to be turned around. He manfully persisted, finally achieving a movable mouthpiece for the conflagration fighter so that the water stream could play in any direction.

What's more, Heron wrote various books describing his inventions in minute detail, with lots of diagrams, pointy arrows, and "put tab A into slot B" instructions. His books won international acclaim—or at least they got translated into Arabic (and thus survived into the modern era). Despite the brilliance and scope of Heron's bright ideas using renewable energy, vast numbers of people energetically ignored them. Even the windmill principle got passed over, to be exploited in another century by an Islamic engineer.

On the other hand, with little fanfare (and no actual fans whatsoever), Greeks and Romans around the Mediterranean harnessed the hydropower of their rivers and streams, building watermills in order to grind grain. Archaeological remnants of such structures continue to be found at forts, towns, and private houses. Once recognized as such, it turns out they were all over the place: at the Athenian agora, near Hadrian's Wall in Great Britain, on Janiculum Hill in Rome.

Do you have any idea how difficult and tedious it was to grind grain by hand each day? That's what working-class Greeks and Romans did in order to make bread or porridge, their mainstay fare. Women often used mortars and pestles for the job. Soldiers in the Roman legions actually carried portable grain mills for their daily grind. What a blessing it must have been to be able to obtain flour from watermill-ground grain.

One particular superlative, a 30-horsepower flour mill near ancient Arelate (modern Arles, France), didn't even rate a passing mention in the books of most ancient writers. Yet this complex of watermills, fed by an aqueduct, used twin millraces and sixteen overshot waterwheels to grind an estimated two to three tons of grain a day. That output could probably have given a town of ten thousand people their daily bread. Green-energy enthusiasts can still admire the sturdy stone walls of this paragon, along with the water channels of its aqueduct.

MAD FOR MORE THAN GADGETS

An early Edison who lived between A.D. 10 and 70, Heron couldn't help himself—the guy was hooked on inventing things, dreaming up wild ideas and seeing them through. His precise descriptions and how-to diagrams for at least seventy-six of his creations are still around. But Heron had versatility—the man was equally adept at penning a mathematical treatise on pneumatics or picking apart theories about a vacuum.

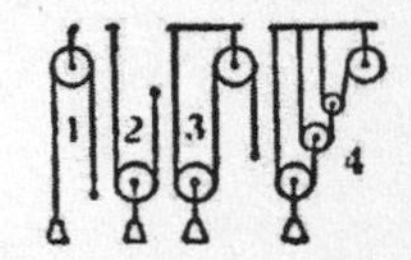

Besides teaching and inventing, Heron wrote a book on mechanics describing the pulley and other key machines. Thankfully it survived.

In some ways, his exceptional career epitomizes the invisible but profound Jekyll-and-Hyde attitudes of his society toward science and applied technology. That split is nicely illustrated by the Roman toga: a useless garment, impossible to put on or keep in place. That was its point. The toga shouted, "This is a man whose clean hands never touch manual labor! A man with inferiors who dress him, who keep that 20 yards of wool clean!" (Like elite Roman males, upper-class Greeks also wore a badge of rank, a bedsheet-sized swath of pristine white wool called the himation.)

It wasn't just physical labor, the down-and-dirty slave chores that old-school aristocrats disparaged; they sneered at any sort of handiwork, from sculpture to surgery. Plato and other philosophers argued that even if a slave were to invent something, he wouldn't own it—that was the province of the aristocratic user, Plato and his ilk.

Thankfully, Heron lived in a less hidebound city: Alexandria, the finest of Alexander the Great's planned cities, with a large, ethnically diverse population.

By Heron's day, the prestigious museum and library, the world's first university and research institution, had clocked three centuries. He is thought to have been a lecturer there; in fact, his extant book on mechanics resembles lecture notes, adding credence to the evidence that Heron taught. His book reveals the theory behind practical matters, such as his description of five machines that, when used, allowed a given force to move a greater weight. The lineup—lever, screw, wedge, pulley, and wheel plus axle—are all key elements used today.

Among the inventions chalked up by this whirlwind Alexandrian are a highly useful surveying instrument called the diopter, a lineup of war weapons including a powerful crossbow, and a device for cutting screw threads into wood. But the invention that has drawn countless sighs of "what if" through the centuries is Heron's steam turbine engine, which he called the aeolopile or wind ball. The bottom half looked like a three-legged sealed kettle of water atop a charcoal fire. Twin pipes connected the kettle with a hollow metal ball on top; the steam caused the ball's two jets to revolve.

The aeolopile was a novelty that (as far as we know) never served a useful purpose. Those who've studied it affirm that the prototype, with its boiler that could raise a head of steam, and his prior use of workable valves and cylinders and pistons that fit, had all the elements needed for a full-blown steam engine. Why didn't Heron's invention find acceptance? Opinions are many. Some blame lack of demand due to reliance on slavery, others the difficulty of obtaining enough high-quality iron for its manufacture; still others cite the lack of high-grade fossil fuel to fire it up. All valid, but perhaps it could also have been the disconnect between the inventive Heron and the times he lived in.

Heron wasn't the only mechanical genius who lived too early for his breakthrough ideas and inventions to realize their full potential. Men such as Ctesibius, Philo, and Archimedes had similar capabilities, and perhaps similar frustrations.

Heron himself drew on their traditions, poring over their writings while developing his own ideas—and at times refining theirs. His earliest source may have been Ctesibius, a local boy three centuries before Heron who wrote now-vanished treatises on his work. His key achievements included workable prototypes of the force pump, the suction pump, the cylinder and plunger, and the parastatic water clock. For the last, Ctesibius was sharp enough to realize that he needed to make the opening of gold or gemstone, something that would not get corroded by the water's action. He also invented a catapult that worked on compressed air. Like other mechanicians, he got his fattest commissions for trivial things, one being a singing cornucopia for the dead wife of the reigning ruler.

A few decades after Ctesibius, a Greek military engineer on the eastern side of the Mediterranean, Philo of Byzantium by name, made it into the engineering hall of fame with a useful device of five interconnected pulleys. He saw it as a good method to work a bucket chain that would haul water. Others used it as a springboard to develop treadmill cranes for building purposes. A genius with pneumatics, he, like Heron and Ctesibius, tinkered with a variety of siphons. His goofy toys included a bird-and-snake affair powered by water. Much of his book, called *Mechanical Syntaxis*, has disappeared, but the portions on pneumatic phenomena, machines, and artillery machines still exist.

A great many of Heron's insights and inventions were put to use in frivolous ways as well: as toys for the wealthy to amuse their guests, and as devices to impress visitors and pull off "miracles" at religious temples. As noted elsewhere, Heron dreamed up the world's first vending machine, which dispensed holy water upon the deposit of a silver coin. His detailed blueprint of it still delights.

Besides genius, unquenchable scientific curiosity, and the opportunity to experiment, what these men had in common was a deep understanding of the mathematical and mechanical principles behind their work. That and, chances are, a secret dislike of clean togas.

GREEK FIRE

Oh, those quirky, absentminded Greeks and Romans. Over the centuries, they had a habit of inventing spectacular breakthrough devices and formulas—then using them for trivial pursuits. Or not at all.

Classic example: the fabled "Greek fire," supposedly a seventh-century Byzantine-era weapon of mass destruction. Modern write-ups, online and off, take a wide-eyed approach to Greek fire, usually stressing its "mystery" ingredients and its "now lost" formula.

On the contrary, careful reading of the accounts left by Greek and Latin sources show that an enigmatic mixture that caused flames to burn intensely even when submersed in water was known in much earlier times.

Unquenchable fire was long used during women's festivals, for instance. Although societal pressure leaned on Greek and Roman married women to maintain matronly decorum, their festivals celebrating the wine god Dionysius (Bacchus in Latin) were periodic escape valves. At these riotous wine-soaked events, married gals became Maenads or Bacchae, followers of the god. Grabbing up torches, the women ran in packs through wooded hillsides all night long—with occasional breaks to make opportunistic kills of wildlife and devour it raw. Whether you called him Bacchus or Dionysius, the god apparently liked to be toasted with steak tartare as well as wine.

Those torches were not your ordinary flambeaux. As the historian Livy, a contemporary of Emperor Octavian Augustus, noted, "The matrons dressed as Bacchae rushed down to the Tiber River with their burning torches, plunged them into the water and drew them out again, the flame undiminished, as they were made of sulphur and pitch mixed with lime."

More than one slick-talking magician made his fortune in ancient times with the showy use of such materials. Around A.D. 200, a fellow who called himself Xenophon used self-igniting fire to dazzle the crowd.

Both Greek and Roman military forces employed arrows impregnated with petroleum mixtures and tinkered with them for centuries to get the most invidious results. By late Roman times, the most popular missile was an arrow coated with a mixture of resin, naphtha, sulfur, salt, and quicklime that would ignite violently on contact with water. An equally devastating alternative was to place those same ingredients inside an airtight bronze vessel, then lob it into wells on land or into seawater near enemy warships.

So the Greek fire of late Byzantine fame was merely a refinement of the incendiary weapon already in almost everyone's arsenal. The man pointed to as the creator of Greek fire, an architect named Callinicus, was a dandy promoter—and an avid borrower. As he burrowed into the literature of earlier mechanical lore, he ran across an invention from what to him was ancient times—200 B.C.—and thought: "This might work." The invention, a piston and cylinder pump, had come from the Alexandrian workshop of a Greek tinkerer named Ctesibius, mentioned elsewhere in this book.

Centuries before Greek fire, revelers at the wine god's all-nighter festivals brandished unquenchable torches made with sulfur, pitch, and lime.

Centuries after Ctesibius, another Alexandrian gadgeteer named Heron had thought the device might make a good instrument for injecting liquids. An enema, perhaps. Or, better yet, a tool for pulling out pus! No one bought it. Still later, Heron wrote a book on warfare; in it, he proposed a larger version of the pus-pulling syringe, repurposed as a flamethrower. This idea didn't catch fire, either.

Reading the accounts of Heron and Ctesibius, Callinicus saw big potential. A possible vehicle to propel his volatile brew? Why not? Building a heavy-duty prototype, he mounted it on the prow of a warship and filled it—verrrry carefully—with the petroleum-sulfur-quicklime combo. One giant *ka-pow* and waterborne conflagration later, the temporarily deafened architect saw that he'd stumbled on a truly nasty new delivery system for WMDs. And before long, a buyer: the Byzantine fleet of Emperor Justinian. Once installed, the Greek-fire-laden ships created barbecues across the water and on enemy ships, even at a distance.

In a nanosecond, Byzantine bureaucrats sealed all access to information about Greek fire and forbade anyone to talk about it. Their efforts at top secrecy were unavailing, however: a journalist spilled the beans, setting off a round of combustible wars between various factions in the Middle East—whose fires, Greek or not, still seem to be unquenchable.

IT HEALS, IT MADDENS, IT MELLIFIES CORPSES

In days of yore, honey got a lot of buzz—with good reason. It was the only sweetener around; boiled-down grape juice ran a distant second. Bee by-products were highly useful; as early as 594 B.C., legislators in Athens had regulations in place for wax and honey. More expensive than honey, wax was critical for everyday products from writing tablets to the casting of molten metals via the lost wax method.

For its cure rate on burns and wounds, honey (and its ability to shrink bacterial cells with antibacterial enzymes) beat out other leading remedies, like dung and rust. The Egyptians, long sold on its antiseptic powers, had nine hundred remedies containing the sticky golden cure-all. Shrewd Greek doctors followed their methods and treated abscesses with honey in which bees had died. Honey-soaked bandages became routine in the Roman army. The sweet liquid also played a frequent if messy role in vaginal contraceptives.

For being so bee-conscious, the Greeks had weird ideas about the insect, believing each hive had a king rather than a queen bee. Furthermore, thanks to Aristotle's wrongheaded insistence, Greeks and Romans alike were absolutely convinced that honeybees sprang from decay. To create more bees, take a young ox into a small house, stop up its nostrils and mouth, bludgeon it to death, and leave it to rot in the room, with thyme and cassia. Voilà! A disgusting mess *and* your very own swarm of honey producers.

On occasion, bees pulled a fast one and sucked nectar toxic to humans from certain species of rhododendron, oleander, laurel, and azalea. The output from such flowering plants was called mad honey. It tasted fine going down but soon victims experienced intoxication, blurred vision, and loss of muscle control that lasted for days. (Modern researchers have studied mad honey, which still afflicts the unwary; its poisonous compounds include grayanotoxin, which inhibits breathing and acts on the nervous system and heart.) In 401 B.C., Greek troops

led by the Athenian general Xenophon got lost in the wilds of Armenia; while there, they bumped into hives of wild honey, feasted on it, and collapsed by the thousands. In 67 B.C., Rome's Pompey the Great was tracking a crafty opponent who put toxic honeycombs in his path. Once Pompey's army was rendered helpless, they were attacked and slaughtered. Pompey clearly was not so Great at geography, since the massacre occurred in the same wilds where Xenophon's troops had gotten wasted.

Mad honey wasn't entirely bad. Devil-may-care drinkers used it to spike their mead or added just a tad to wine. And, as noted folklorist Adrienne Mayor suggests, women who prophesied at oracles like Delphi or engaged in Dionysiac religious rites like the Maenads may have nipped at mad honey to become divinely maddened.

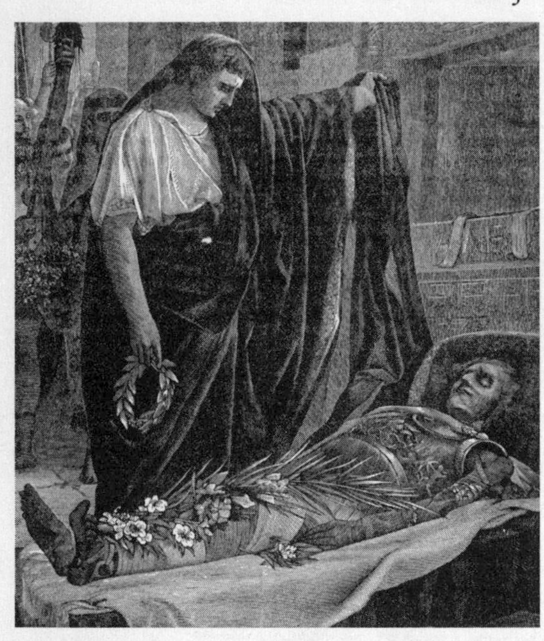

Julius Caesar paid his respects to Alexander the Great—280 years after the Great one had died.

Then as now, honey's invaluable qualities far outweighed its infrequent dangers. Another of its properties saw quite a lot of use in ancient times: mellifying.

It had long been common knowledge that the Babylonians embalmed with wax and honey. But the big news began when Alexander the Great died at age thirty-three. Always organized, Alex had left pre-need instructions to mellify his remains. The high sugar content of honey draws water from cells and gradually dehydrates tissues. Thus, if honey happens to surround a corpse, under the right conditions it produces a drying action while also preserving. It seemed to work for Alex. His body survived a 1,000-mile road trip, a corpse-napping, and decades-long display in a glass coffin in Memphis, Egypt—and he was still being called 'lifelike' when last seen centuries later by Roman Emperor Caracalla.

Although honey has been used for millennia to preserve fruit and meat, the length of time it can preserve a human body has long been the subject of debate. In a relatively calm interval that the Iraqis enjoyed between recent wars, scientists at the College of Medicine in Baghdad, Iraq, conducted a seven-year study using mice, rabbits, and two human fetuses. Once viscera were removed, the cadavers were embalmed in unprocessed honey for a month, then dried and kept in closed glass boxes at room temperature for one to three years. The results? Subjects shrank and mummified while retaining their shape—no decay at all.

In addition, the study confirmed the viability of preserving skin grafts in this fashion. Skin biopsies kept in honey dehydrated; on transfer to saline solutions, they regained their consistency, texture, and color.

Will honey recover its ancient allure and become the wonder drug of the twenty-first century? It's possible. Since it kills bacteria in a different way than drugs like penicillin do, it's being evaluated as a weapon against the growing number of drug-resistant germs.

Honey could rock the funeral business as well. Today's embalming chemicals are largely cosmetic, achieving a transient rosy-cheeked appearance to please the bereaved. Once a casket goes underground and moisture hits its occupant, decay and mold set in with vigor. Honey, the organic alternative to conventional embalming? It could work—that is, if pesticide overuse, climate change, colony collapse disorder, and other threats have left any bees to make the sweet stuff.

FINAL LAUNCHES TO REMEMBER

Death rites had high priority among those still aboveground. Neither Greeks nor Romans were sanguine about the prospects of a rewarding afterlife. The final launch, however, had to be done right because ghosts were easily enraged by corner cutting. Besides, the deaths of conspicuous citizens promised spectacle and largesse. Funerals were community affairs where many came to pay respects and collect goodie bags, traditionally filled with raw meat. Since even those of modest means were meticulously memorialized, the funerals of the great or even the quasi-great had to be spectacular.

Grandiose funerals? Mandatory for the rich and famous. Seen less often were the DIY variety—modest but undeniably spectacular.

For instance, take that autumn day in 324 B.C. when Hephaestion, intimate best friend of Alexander the Great, died of fever. Since they happened to be in Mesopotamia (present-day Iraq) at the time, grieving Alex designed his friend's funeral pyre to resemble a Babylonian ziggurat, with seven wooden stages depicting battle scenes.

The bottom deck of this monstrosity displayed the bronze prows from 240 war galleys, flanked by alternating rows of archers and hoplite soldiers waving red felt banners. At the top of the ziggurat, oversized images of the seductive sirens from *The Odyssey* sang funeral laments, the warbling supplied by human singers hidden inside. (Before the pyre was fired up, presumably the huge cast of humans was allowed to climb down.) The tab for the whole schmear was said to be 12,000 talents—a sum impossible to calculate in today's terms, but in the low gazillions, without a doubt.

Hephaestion's wake foreshadowed a greater tragedy: the sudden demise of Alexander himself in the spring of 323 B.C. Shortly before his death, Alex got trapped for several days aboard a boat in a mosquito-rich marsh near Babylon. After the rescue, he followed this fact-finding expedition with a series of hard-drinking parties in late May. Alex fell ill, lingering in a high fever until June 10, when he was pronounced dead, according to written eyewitness accounts. The Babylonian embalmers arrived on June 16, expecting a real mess. Despite the intense summer heat, Alexander's body had not decomposed.

That eerie fact has motivated modern medical researchers to research the possibilities. They hypothesize two plausible scenarios. One is terminal coma, a fairly frequent complication of malaria from the *P. falciparum* parasite. Typhoid fever may also present a complication called ascending paralysis, a neurological problem that moves up the patient's body and gives the appearance of death.

When finally given the go-ahead to embalm, the Babylonians applied the honey mellifying treatment requested by Alexander and (we're guessing here) some tricks of their own. Whatever their methodology, as you'll see, Babylonian honey embalming was worth every talent it cost.

Alexander's still-pristine corpse sat around for nearly two years while his successors fought over territory and artisans labored to make an exquisite coffin of beaten gold. After his successor generals jointly decided to carry their departed leader through the lands he'd conquered to his Macedonian birthplace, they commissioned an elaborate carriage. Drawn by sixty-four mules, topped by a golden temple, his blinding coffin was guarded by life-sized lions of precious metal. The funeral cortege moved from city to city, crossing a thousand miles of Asia. It attracted huge crowds, some following the golden coffin for days.

When it reached Syria, General Ptolemy, one of Alex's inner circle, came with his army to meet the cortege. In the words of author Mary Renault, "Ptolemy's homage was to carry out a reverent hijack." Instead of going to Macedonia, the cortege now headed south to Egypt, where Ptolemy parked the dear departed in the city of Memphis for another decade or two. This was but a pit stop; Ptolemy already had crews working on what he and the Alexandrians called the Tomb.

About 215 B.C., the vagabonding corpse of Alexander the Great was finally installed at the Soma (Greek for "the body"), a lavish pyramid-shaped mausoleum in Alexandria. The complex, featuring an underground chamber for Alex, quickly gained a touristic following.

After the first Ptolemy died, his dynasty gradually grew weaker, culminating in the financially strapped Ptolemy IX, who melted down Alex's golden sarcophagus to make coinage for his mercenary army. The outraged Alexandrians murdered Number Nine in due course, meanwhile confecting a new sarcophagus made of colored glass.

According to all accounts, Alexander still looked Great. The Soma continued to attract crowds, including Julius Caesar. The ultimate Caesar to lay eyes on Alex may have been Emperor Caracalla, who sealed the chamber for the last time in A.D. 215. After that, turbulent politics and the sack of Alexandria destroyed the above-ground structure. A huge earthquake and tidal wave in A.D. 365 did the rest.

From the sublime to the ridiculous—we turn now to the funeral pomp of one Pythionice. Ring a bell? Didn't think so. A contemporary of Alexander's inner circle, she was considered the hottest hetera going. Harpalus, a rascally childhood buddy of Alex who became his disastrous pick as army paymaster, fell hard for her. After some dubious dealings, including large-scale embezzlement from the riches amassed by Alexander and his armies, Harpalus brought Pythionice to Babylon, lavishing gifts on her and treating her like a queen. After having Harpalus' first child, Pythionice had the nerve to die on him.

Not one to hold a grudge, Harpalus proceeded to memorialize Pythionice in the most maximum way ever accorded an ancient Greek woman. Not content with a miserly one-tomb concept, he erected a magnificent pair of matching stone memorials to her at Eleusis, near Athens, and in Babylon itself. For the funeral procession, he escorted her body, accompanied by a huge processon of top-notch musicians and singers, along the Sacred Way between Athens and Eleusis.

The funereal blowout cost in the neighborhood of 200 talents—chicken feed compared to the splash for Hephaestion, but still impressive. Harpalus, however, was on speed mourning. Not long after the wake, he took up with Glycera, another top-ranking hetera.

But nothing could take away the lasting impact of that tall roadside attraction: a pyramid- or wedding-cake-shaped structure with its own altar and temple fondly dedicated by Harpalus to his make-believe goddess, Aphrodite Pythionice. Nearly five hundred years later, that intrepid geographer and travel writer Pausanias visited it. In his maddeningly prosaic prose, he utterly failed to describe the shape, size, or specifics of Pythionice's resting place. Instead, in his plodding prose he gave it a four-star recommendation as "the best worth seeing of all ancient Greek tombs."

INVENTING GEOGRAPHY

Cartographers and geographers of old had countless shouting matches over what constituted their world. Some described it as round: Greece in the center, the Oracle of Delphi as the belly-button. Others, like the normally sensible Democritus of Abdera, swore that the earth was rectangular. Working independently, two math geeks came within an ace of nailing the correct circumference of the globe. Nearly everyone else, however, bought into an opposing theory from an influential big noise called Claudius Ptolemy, who insisted that the earth was half its actual size.

Only one thing did this quarrelsome bunch agree upon: the world's three continents of Europe, Asia, and Libya (the original name given to Africa) were completely surrounded by waters so vast they were simply designated Oceanus.

While scientists and armchair jockeys squabbled, a handful of intrepid Phoenicians, Greeks, and Romans set out to actually explore the globe on foot and on water. Bits and pieces of their firsthand reports are still with us, although officials of long ago made scant use of the geographic gold they produced.

Looking at a modern map, the quasi-closed system of the Mediterranean Sea, with its deeply corrugated coastlines, its plethora of islands, looks like a snap to navigate and explore. It wasn't. The same went for the Black Sea, named Euxine or "good waters" by the Greeks, who used that term in the same way we insincerely mutter "Good doggie!" to a snarling pit bull.

The ships that explorers used weren't especially safe or seaworthy. "When in doubt, run for land" was every captain's motto. Lacking compasses and coastal maps, most vessels were at the mercy of pirate vessels and rough weather. Shipwrecks were as common as pileups on California freeways. From October to March, the sailing season slammed shut.

The Phoenicians of Carthage, the ancient world's slickest traders, took the first steps toward a more fact-based geography. Having already explored, colonized, and built a customer base around the Med, they hankered to get a handle on the African continent.

Their maiden voyage was a bold one. In the seventh century B.C., they formed a joint venture with Necho, an Egyptian pharoah of the twenty-sixth dynasty, who dug the first canal between the Nile and the Red Sea, then helped subsidize a three-year east-to-west voyage around Africa. By happy coincidence, that happened to be the most favorable direction to catch currents and winds. The Phoenicians did indeed circumnavigate Africa—but afterward, their reports of seeing the sun on their right side were snickered at as tall tales, and largely forgotten.

Around 500 B.C., a forceful John Wayne sort of expedition leader named Hanno tackled another mission, sponsored by the Carthage Colony-Planting Task Force. With sixty well-provisioned ships jammed with colonists, he sailed through the Pillars of Hercules (the Straits of Gibraltar today) and down the west coast of Africa. After offloading colonists in promising parts of Morocco, he sailed on, encountering savage tribes, the first big apes ever seen (promptly killed and their skins made into trophies for Carthage's Temple of Astarte), and an active volcano called the Seat of the Gods. Not a spendthrift with words, upon his return Hanno wrote an eighteen-sentence account of his voyage, putting it on a bronze tablet in a Carthage temple. His pithy summary was later copied and translated by the Greeks.

One of the most adventurous explorers to get up a creek without a paddle was Pytheas, a geographer from Massilia (present-day Marseilles, France).

Like the Phoenicians, the Massilians had a talent for travel and a keen eye for new resources. Pytheas' trip may have been funded by traders hoping to find more sources for tin.

Pytheas sailed solo from his home, hung a right at Portugal into the Atlantic Ocean, and circumnavigated Britain. He hiked that island, describing one marvel that sounded a lot like Stonehenge, which he said was surrounded by a city worshiping a sun god. Back on board, Pytheas headed for what he called "the ends of the world." After sailing north for six days or more, he saw magical sights: the sun shining at midnight, frozen seas in the Arctic circle, and a place he dubbed Ultima Thule. It's thought he may have reached Iceland and/or Norway.

When Pytheas got safely home, he reckoned he'd sailed about 7,500 miles—the equivalent of Columbus' first voyage. Around 320 B.C., he wrote about his travels. His book, *On the Ocean*, with its priceless insights about physical geography and northern European cultures, has only survived as tatters within the writings of Strabo, Polybius, and other authors.

The literary output of explorer Arrian, while duller, had more luck. As Roman governor of the frontier province of Cappadocia, he made an arduous circuit of the Black Sea and wrote *Periplus Ponti Euxini*, a coastal guidebook. Strabo, a Greek from the same part of the world, had similar good fortune. His seventeen books on geography survived, along with now-lost information on the mathematical work of scientists like Hipparchus and Eratosthenes, the latter being one of the two men who calculated the nearly exact circumference of the earth.

Even as geography became clearer, travel itself remained hazardous. That didn't stop less rugged types from globetrotting. In the reign of the first Roman emperor, an obscure philosopher named Potamon set out for foreign lands, armed only with a piece of papyrus. His safe-conduct document had a ferocious, ahoy matey tone: "If there be anyone on land or sea hardy enough to molest Potamon, let him consider whether he be strong enough to wage war with Caesar."

For sheer persistence, the dogged traveler-explorer award must go to Etheria, an early Christian nun with a secret yen to globe trot. In late Roman times, she undertook a three-year religious pilgrimage from Spain to the Holy Land to Constantinople to Egypt and back again, writing in her diary every bug-ridden, bandit-infested step of the way. The only thing worse than rigors of the trip itself? Being forced to read her colorless telling of it. Perhaps Etheria had an ulterior motive for making her narrative deadly dull. Once the other nuns she lived with caught a glimpse of the wild and wonderful wider world, how would anyone keep them down in the nunnery?

FORGET THE GARDENS, BUILD ME A LABYRINTH

Not much existed in the way of natural wonders around the Mediterranean Sea two thousand years ago—no Grand Canyon, no Niagara Falls. Instead, travelers drooled over exquisite artworks. Or gasped at big, impossibly difficult-to-build things—like the pyramids of Egypt.

From century to century, the ancient world wonders list varied but usually included: the pyramids and labyrinth of Egypt; the Mausoleum at Halicarnassus; the Artemis temple at Ephesus; the hanging gardens and walls of Babylon; the lighthouse at Alexandria; the colossal statue on Rhodes; the statue of Zeus at Olympia.

Despite strenuous counterclaims from the Greeks, Romans, and Babylonians, Egyptian wonders won the most plaudits. From the Sphinx to the obelisks, they oozed power, authority, and beauty. The labyrinth on Crete having long since vanished, visitors could head for Egypt's version. Visited by Herodotus in the fifth century B.C. and geographer Strabo five hundred years later, the site elicited confused wonder. Neither man was allowed to visit the subterranean labyrinth, its chambers allegedly numbering three thousand. The thousands of shadowy rooms above ground were labyrinthine too, a series of roofed courts filled with pillars and gorgeous paintings and hieroglyphs, a confusion of corridors and myriad doors, real and fake.

Pliny, who never laid eyes on the labyrinth, seemed bewitched by it. As he babbled, "Quite the most abnormal achievement on which man has spent his resources, but by no means a fictitious one . . . there is no doubt that Daedalus adopted it as the model for the labyrinth built by him in Crete, but he reproduced only a hundredth part of it." Pliny goes on to say that there was a third labyrinth in Lemnos and a fourth in Italy.

About 1888, archaeological pioneer Flinders Petrie found a gigantic rectangle of worked stone measuring 798,597 square feet south of the Hawara pyramid. He called it the labyrinth site but felt the aboveground structure itself had collapsed. A decade or two later, Egyptian engineers began to build a railway in the area; sadly, they cannibalized much of the white limestone paving on the labyrinth site for their train roadbed.

In 2008, a joint Egyptian-Belgian expedition used advanced ground-penetrating techniques to scan the area. Results may soon provide some answers—and determine if any of the labyrinth's underground structures and grid are intact.

Only marble crumbs remain of another ancient wonder: the Mausoleum at Halicarnassus (modern Bodrum, Turkey). Once resembling a many-tiered cake with wildly colored icing, the 140-foot-tall building was begun by King Mausolus about 360 B.C., then completed as his tomb by his grieving wife, Artemisia II. (These slipshod days, any old crypt gets called a mausoleum.)

The Mausoleum sat at the crossroads of the city's main streets, a landmark for ships at sea. For the niches between its columns, five men judged as the world's best sculptors crafted more than a hundred colossal figures. This wonder stood for fifteen hundred years until that fateful day when the crusading Knights of St. John cruised by, spotted its acres of fine marble, and gleefully began to dismantle it for their fortress nearby—where their disappointing efforts can still be seen.

Given the difficulty of long-ago travel, few travelers got as far as Babylon, fabled home of two world wonders: its city gates and its hanging gardens. Glittering Babylon, whose name still scintillates, a city of legend. Literally, as it turns out.

Its city walls, thrillingly described as wide enough for two chariots to race each other on top, were said to stand 75 feet high and circle the city for 40 miles or more, with one hundred gates of bronze. (The measurements given by the Greek historian Herodotus are even more fanciful, but at least he never claimed to have actually seen them. Or been there.) In modern times, Iraq's former dictator, Saddam Hussein, installed a fake Ishtar Gate and walls on top of what's left of the originals, so it's unlikely we'll learn any time soon what they really looked like.

Ah, but the hanging gardens of Babylon. Judging by the breathless accounts of Philo, Cleitarchus et alii, their terraced tiers of greenery stood as tall as those city walls. They held a forest of mature trees, with stone pillars below to support the structure; ingenious machinery handled watering and drainage. To date, nothing resembling the hanging gardens has ever been found. Zip. German archaeologist Robert Koldewey excavated here for two decades, finally encountering stone pillars which he proclaimed were the gardens' base. They turned out to be prosaic buildings for Babylonian bureaucrats, too far from the river to feasibly water houseplants, much less vertical gardens.

Where there's a wonder so wondrous it's too good to be true, it probably isn't. And wasn't. But in that ferocious climate, who wouldn't begin to hallucinate dreamy, luscious gardens and well-watered greenery?

The prize, however, for the most exquisite wonder no longer with us goes to the seated statue of Zeus at Olympia. Three stories high, made in the style known as chryselephantine, the noble, brooding figure was carved of wood, then overlaid with ivory and gold, the god's scepter and crown encrusted with jewels. Like the 50-foot statue of Athena he'd made for the Parthenon, Pheidias the sculptor planned the relative size of the statues and the buildings containing them for maximum impact.

When he'd finished, the sculptor casually asked Zeus if it pleased. The god responded with a thunderbolt, splitting the pavement at Pheidias' feet. Unveiled in 438 B.C., the Zeus drew crowds at the following year's Olympics. The adoring mob was kept at arm's length by a pool of olive oil—thought to be needed to preserve the artwork's ivory from cracking. Pheidias' extraordinary work survived for eight centuries in its sanctuary. When Rome split in two, his Zeus statue was snatched up by Byzantine bigwigs, ultimately disappearing in the flames of the Constantinople fire of A.D. 475.

Beyond the official world wonders, there were plenty of fantastic also-rans to draw pilgrims and impress travelers. The Parthenon and Acropolis of Athens. Rome's Pantheon and Colosseum. The gigantic Temple of Jupiter at Baalbek. Alexandria's Great Museum and Library. Travelers with the stamina for it could hit the entire circuit of wondrous cities, shrines, and monuments. Delphi was big. Olympia was big. Mt. Etna was huge. So was Troy, in the other sense of the word. Places with the most mythological and historic tie-ins, authentic or imagined, often got the biggest raves.

The word *awe* has feeble firepower these days, but when it was newly minted, it conveyed an almost fearful reverence, a holy wonder at something sublime. That's a quality that ancient folks, with plenty to be awed at yet thrilled by a magnet, had in abundance.

ACKNOWLEDGMENTS & SPECIAL THANKS

Warm thanks to my agent, David Forrer, and all the terrific folks at InkWell Management; and to my dream publisher, George Gibson, and the wonderful staff at Walker/Bloomsbury, especially my editor, Margaret Maloney; production editor, Laura Phillips; my keen-eyed copy editor, Susan Warga; proofreader, Nancy Gilbert; and publicist, Sara Mercurio.

My gratitude as well to ace researchers Sharon Morem and Kim Dunn, and to multitasking cartographers Margaret Maloney and Ashala Lawler. Finally, my heartfelt thanks to the following circles of friends, fellow authors, librarians, educators, and generous experts who gave of their time and knowledge to review material, give feedback, and offer opinions:

Chet Amyx; Ellen and Gilbert Ansolabehere; Richard Blake; Dr. Jeffrey Bloom; Dorothy Buhrman, George Burns, and all the members of the Cambria Writers Group; Val and Kevin Conroy; Lucia and Stephen Davies; Luciano and Maria Grazia di Dio; Ruth Downie; Doug Dunn; Olga Essex; Richard Ferraro; and Margot Silk Forrest.

Jim and Joan Griffin; Grant and Kate Gullickson; Michael Haag; Jim Hayes; Jack Harris; Judith Harris and David Willey; Mary Harris; Caroline Hatton; Shera Hill; Carolyn Hornbuckle; and Joni Hunt.

Laura Keefe; Douglas Kenning; Richard Klein; Dan and Liz Krieger; Barbara Lane; Brian Lawler; Stephanie Lile; Jude Long and the staff of the Morro Bay Library; Maria Lorca; Jim Loring; David Loring; Sheila and David Lyons.

Robin Maxwell and Max Thomas; Philip Manor; Michael May; Bill Morem; David and John Moore; Tom Neuhaus; the Ogren clan; Mark Phillips; Dorothy Pier; and Dean Poysky.

Bob and Martha Raaka; Cindy and Curt Rankin; Jeri Remley; Sherry Shahan; John Siscoe; Elizabeth Spurr; Stephanie Spurr; Xiaoping Shanbrom; Diane Stevens; Judy Sullivan; Diane Urbani de la Paz; Tom and Elizabeth Wayland-Seal; Wayne Wilson; Richard Wortman; Joyce and Phil Wyels.

BIBLIOGRAPHY

Our Greco-Roman legacy provides any researcher with an embarrassment of riches. Valuable primary sources include Athenaeus; the *Res Gestae* of Emperor Octavian Augustus; Cicero's *Of Divination* and his own letters; Diogenes Laertius's *Lives of Eminent Philosophers;* Diodorus Siculus; Epictetus; the Landmark editions of Herodotus and Thucydides; Iamblichus; Livy; Marcus Aurelius' *Meditations;* Pausanias' *Description of Greece;* Phlegon's *Book of Marvels;* the letters and works of Pliny the Elder and the Younger; Porphyry; Plutarch's *Moralia;* Seneca; Strabo's *Geography;* Tacitus; Theophrastus' *Inquiry into Plants;* and Vitruvius.

Other priceless material comes from ancient coins, period artwork, and a cornucopia of archaeological sources; from primary and secondary sources online, notably William Thayer's vast chef d'oeuvre LacusCurtius; and from letters, inscriptions, and primary-source compilations found in books such as Shelton's *As the Romans Did,* Guthrie's *Pythagorean Sourcebook,* and Hunt's *Select Papyri.*

Explore the following for in-depth looks at unusual aspects of ancient science or superstition:

Beard, Mary. *The Parthenon.* (Profile Books, 2002.)

Casson, Lionel. *The Ancient Mariners,* 2nd ed. (Princeton University Press, 1991.)

Cohen, Morris, and I. E. Drabkin. *A Source Book in Greek Science.* (McGraw-Hill, 1948.)

Cunliffe, Barry. *The Extraordinary Voyage of Pytheas the Greek.* (Walker & Company, 2002.)

Felton, D. *Haunted Greece and Rome, Ghost Stories from Classical Antiquity.* (Texas University Press, 1999.)

Forbes, R. J. *Studies in Ancient Technology, 8 vols.* (E. J. Brill, 1966.)

Hirshfield, Alan. *Eureka Man.* (Walker & Company, 2009.)

Holmyard, E. J. *Alchemy.* (Penguin, 1968.)

Hopkins, Keith, and Mary Beard. *The Colosseum.* (Profile Books, 2005.)

Hughes, J. Donald. *Pan's Travail.* (Johns Hopkins University Press, 1994.)

James, Peter and Nick Thorpe. *Ancient Inventions.* (Ballantine Books, 1994.)

Lewis, Naphtali. *The Interpretation of Dreams & Portents in Antiquity.* (Bolchazy-Carducci Publishing, 1996 reprint.)

Luck, Georg. *Arcana Mundi.* (Johns Hopkins University Press, 2006.)

Mayor, Adrienne. *Greek Fire, Poison Arrows & Scorpion Bombs.* (Overlook Duckworth, London 2004.)

____. *The First Fossil Hunters, Paleontology in Greek and Roman Times.* (Princeton University Press, 2000.)

Moore, David. *The Roman Pantheon: the Triumph of Concrete.* (Office Outlet Publishing, 1995.)

Pedersen, O., and Mogens Pihl. *Early Physics and Astronomy.* (Macdonald and Janes, London 1974.)

Smith, William. *Dictionary of Greek and Roman Antiquities,* 2nd ed. (Little, Brown, 1859.) Most of the Roman entries are also available, searchable, and annotated with acerbic wit online at Bill Thayer's LacusCurtius Web site.

Strong, Donald, and David Brown, eds. *Roman Crafts.* (Duckworth, London 1976.)

Temple, Robert. *The Crystal Sun.* (Century Books, London 2000.)

White, K. D. *Greek and Roman Technology.* (Cornell University Press, 1984.)

Woodcroft, Bennet, translator, ed. *The Pneumatics of Hero of Alexandria.* (online at www.history.rochester.edu/steam-hero/index.html)

INDEX

F

G

H

I

L

A NOTE ON THE AUTHOR

Author of more than thirty nonfiction books, including the popular Uppity Women in history series, Vicki León wanders widely but calls coastal California home.

"As a writer and researcher, I often journey back in time as well as vagabonding through geographic space. The graceful restlessness of the triskelion is my talisman, a good luck piece to return me to my original starting point, to begin anew."

An ancient symbol of good fortune, the triskelion originally represented the circular movements of the life-bringing sun, its swift feet echoing the journey through life experienced by all creatures.